建筑立场系列丛书

建筑的公共与私域

ARCHITECTURE

Pseudo-public Pseudo-private

于风军 罗茜 张可新 陈颖 翟白玉 于承志 韩瑶 孙宁宁 孙艳阳 李楠楠 高一博 | 译

[英]福斯特建筑事务所等 | 编

大连理工大学出版社

建筑的公共与私域

Architecture – Pseudo-public, Pseudo-private

Architecture
Pseudo-public, Pseudo-private

建筑的公共与私域

公共和私有的概念并非完全对立，而是相互交融，相互依存。供个人和集体使用的两种空间之间的交融互动决定了公共建筑的成功与否。本文探讨了公共活动和个人感知体验、公共所有权和私有化空间如何在几个不同的公共建筑类型中展开，例如公共开放空间、公共游泳池、混合用途建筑和博物馆，并通过伦敦、香港、赫尔辛基和哥本哈根等不同城市的建筑实例进行阐述。本文还回顾了富有争

The concepts of public and private are not diametrically opposed but rather exist on a continuum. The interaction between the two states – for the individual and the collective – determines the success of public architecture. The article considers how public activity and private sensations, public ownership and privatisation, play out across several different public architecture typologies – such as public open spaces, public swimming pools, mixed use buildings and museums – and elaborates by drawing on built examples from diverse cities including London, Hong Kong, Helsinki and Copenhagen. The article also

ECTURE

public

private

议且仍僵持不下的关于私有公共空间的辩论：这种私有公共空间快速增长的现象如何形成对公民自由和设计自由的冲击，以及冲击的程度有多大。但不同的是，本文讨论了建筑如何在遏制或改变私有公共空间激增所带来的负面影响方面发挥关键作用。最近以公共建筑为主题的国际建筑节和活动（如2018年威尼斯双年展）上的建筑就证明了这一点。

reviews the contentious ongoing debate about POPS (Privately Owned Public Spaces): how this rapidly accelerating phenomenon can be said to impinge on civil liberties and design freedoms, and to what extent. Conversely, the article discusses how architecture can play a pivotal role in curbing, or redirecting, the negative consequences of the proliferation of POPS, as evidenced by contributions to recent international architectural festivals and events – such as the Venice Biennale 2018 – which have used the theme of Public Architecture as a springboard.

建筑的公共与私域
Architecture – Pseudo-public，Pseudo-private

Michèle Woodger

"公共"绝非一个非常宽泛的术语。本书中的建筑项目从车站广场到体育设施，从市场到博物馆，不一而足。这些项目为各行各业的人而建造，目的是通过举办使生活更加充实的文化和娱乐活动来惠及各行各业。

从某种意义上说，几乎所有建筑都是公共的，因为建筑终要呈现在人们的视野中，这一点无法避免。我们一离开自己的家便将自己置身于"大庭广众"之下；不同程度的隐私以及不同的行为在室内或室外都是可以接受的。但令人想不到的是，在拥挤的咖啡馆里，我们可能比在自己家里有更多的私人空间。建筑评论家伊恩·奈恩曾戏谑地将其当地的酒吧描述为自己的"办公桌"，声称："叽叽喳喳的谈话背景声给人一种真正的私密感。我怀疑几乎没有人知道我今天在哪里，这简直太棒了！"[1]

在现在的数字化环境中，这个概念更是争议多多。在这个由GPS、数据、wi-fi定位、监控录像和无人机组成的崭新世界中，我们要劳力费心地躲避监听监控。而与此同时，社交媒体不断期望我们保持公众形象。用班克斯基颇具调侃的话说："现在默默无闻不是比出名更难了吗？以今天的文化视角来看，确实如此。我不知道为什么人们如此热衷于公开自己私生活的点点滴滴；他们忘记了看不见的东西才是一种巨大力量。"[2]

我们还必须将"公共建筑"与"公共所有权"区分开来。一些国家投资的建筑，如议会、监狱、军事设施等，不是公众可以随意出入的。相反，私人购物中心、咖啡馆或画廊则依赖于公众而存在，并且可能因为公众将之视为"自己的"而充分融入一座城镇中去 。

公共和私有并不是完全对立的，这两个概念相互交融，相互依存。可以说，好的公共建筑能成功地协调两者之间的相互作用，利用两者之间存在的矛盾对立来创造具有独特活力的空间，在提高集体活动质量的同时保持单独空间的完整。

让我们来看一下这些矛盾对立统一的状态是如何在开放的空间中存在发展的。我们期望有很多属于我们自己的广场、集市和公园。我们

"Public" is an impossibly broad term. The projects featured in this book range from station plazas to sports facilities, covered markets to museums. They are built for people of all walks of life to benefit in pursuit of the cultural and convivial activities which make life fulfilling.

In one sense almost all architecture is public, since buildings are unavoidably noticeable. We are all "in public" the moment we leave the house; different levels of privacy, and consequently different behaviours, are acceptable indoors or outside. Perversely, we may have more privacy in a crowded café than we do in our own homes. Architecture critic Ian Nairn once jokingly described his local pub as his "office desk", claiming: "the background buzz of conversation gives a real sense of privacy. I doubt if half a dozen people…know where I am today and that's marvellous."[1]

The concept becomes even more problematic in the digital environment. In this brave new world of GPS, cookies, wi-fi positioning, CCTV and drones, we struggle to escape scrutiny. Meanwhile social media places constant expectations on us to maintain a public persona. In the words of the elusive Banksy: "Is it now harder to cultivate anonymity instead of fame? In today's culture: yes. I don't know why people are so keen to put the details of their private life in public; they forget that invisibility is a superpower."[2]

We must also decouple "public architecture" from "public ownership". Some publicly funded buildings – parliaments, prisons, military facilities, etc. – are necessarily inadmissible. Conversely, private shopping malls, cafés or galleries depend upon the public for their existence, and may be sufficiently embedded in the identity of a town to be embraced by the public as "theirs".

Public vs private is a false dichotomy; the two notions are not mutually exclusive but exist on a continuum. It could be argued that good public architecture is that which successfully navigates the interplay between the two, harnessing the tension which exists between them to create spaces with unique vitality. It enhances the collective experience whilst

占领伦敦抗议示威活动，芬斯伯里广场，2011年
Occupy London Protests, Finsbury Square, 2011

圣保罗大教堂，帕特诺斯特广场
St. Paul's Cathedral and Paternoster Square

期待它们应该具有包容性和可达性，"干净且安全"。它们应该迎合群体，为自发的集体活动提供便利，但同时也得是一方净土。它们应空旷而宽广，但也要有长凳、喷泉、纪念碑和售货摊。

许多开放空间为私人所拥有，这一点有时并非显而易见。这并不新鲜，多亏富人出资修建，才有许多罗马和文艺复兴时期的杰作留存于世。在美国，私有公共空间也已经流行了几十年。

今天，我们越来越依赖私人土地所有者为公共领域提供资金，这种情况已形成全球化规模，并引起热议。去年，《卫报》发布了一份伦敦私有公共空间地图[3]。该地图显示，在这些私有公共空间，公众因无意的违规行为，如在禁拍景点拍照、骑自行车和坐在长凳上睡觉，会被保安逮捕或驱逐。这一行为引起了公众恐慌。（在新加坡，提供私有公共场所的人必须在私有公共建筑上贴上一张刻有管理人员姓名和联系方式的牌子，但伦敦并没有采取类似做法。[4]）一名记者在市长办公室（市政厅）外调查伦敦私有公共空间情况时甚至受到警告。

记者们联系了一些公司的土地所有者，在发出的50份调查中只收到两个回复，他们不合作丝毫没有减轻人们对公民自由的担忧。事实上，许多人认为"占领伦敦"运动就是证据：2011年，由于帕特诺斯特广场是一个私有公共空间，抗议者无法在伦敦证券交易所外举行抗议示威活动。在伦敦老城内，只有两个地方可以举行抗议示威活动：圣保罗大教堂和芬斯伯里广场的一小块草地。

具有讽刺意味的是，城市空间不仅不欢迎人们，还让人望而却步。英国也有一项名为《公共空间保护令》的令人质疑的政策：一些由非民选议会官员通过的令人唏嘘的限制条款，不是迂腐老套，就是怪异荒诞。在一些地方，游荡、成群结队、街头卖艺、出售幸运护身符，甚至把羊带进村子，都为法令所禁止。[5]

用愤世嫉俗的眼光看待许多现代广场的商业主义，也情有可原。这些广场的设计是为了鼓励人们走进邻近的咖啡馆和商店。宁静的绿色

preserving the integrity of the individual.

Let us consider how these paradoxical states play out in open spaces. We expect a lot of our plazas, piazzas and parks. They should be inclusive and accessible, "clean and safe". They should cater for groups, facilitating spontaneous collective activities, but also be places of calm. They are characteristically empty, but also filled with benches, fountains, memorials and market stalls.

It is not immediately obvious that many open spaces are privately owned. This is not new: we have wealthy patrons to thank for many Roman and Renaissance masterpieces, and POPS (Privately Owned Public Spaces) have been common practice in the USA for decades.

Today our increasing reliance on private landowners to bankroll the public realm takes place on a global scale and is becoming polemical. Last year *The Guardian* published a map of POPS in London,[3] and revealed that members of the public had been threatened with arrest or evicted by security guards for unwitting infractions such as photography, cycling and sleeping on benches. (In Singapore, those who provide POPS must affix a notice indicating management name and contact details, but not so in London.[4]) A journalist was even admonished outside the mayor's office (City Hall) while researching the story.

When reporters contacted the corporate landowners they received only two replies out of 50; the lack of cooperation did nothing to allay concerns about civil liberties. Indeed, many claim the "Occupy London" movement as proof: in 2011 protesters were unable to demonstrate outside the London Stock Exchange as Paternoster Square is POPS. Only two zones in the Square Mile remained available: St Paul's cathedral and a small patch of grass at Finsbury Square.

Ironically, municipal spaces can be equally uninviting and prohibitive. The UK also has a questionable policy called PSPOS (Public Space Protection Orders): stifling restrictions passed by unelected council officials which range from the banal to

©Alberto Sinigaglia/LUCA Luxembourg Center for Architecture

2018威尼斯双年展卢森堡展厅
Luxembourg Pavilion at Venice Biennale 2018

环境确保来吃午餐的人买上一杯咖啡喝，等下午回到办公室时，他们的大脑就重新充满电，且充满创造力。

我们也为净化后的公共空间的感到遗憾，它限制了意外事件的发生，排除了麻烦状况。正如赫勒·尤尔在《公共建筑：从熟悉到陌生》一书中写道，一座城市的优势在于其多样性，不同利益和从属关系的人尽管有所不同，但仍能和睦相处。当公共空间被"个人利益和服务自我的考虑所驱使，社会的凝聚力就会消失，它的发展潜力也会随之消失。"[6]如果所有事情都按部就班、一成不变，那才是可怕的。

但是建筑也能起到拯救作用。在卢森堡，只有8%的土地归国家所有。[7]在今年的威尼斯双年展（主题为"自由空间——主题的生动选择"）上，卢森堡展厅的展览"公共场所的建筑"展示了建筑师尚未完成项目的大型中密度纤维板模型。这些建筑项目最大限度地利用了架空建筑下方的空间。例如，勒·柯布西耶未建的Îlot insalubre n°6项目包括幼儿园、电影院和图书馆。Andrea Rumpf馆长和Florian Hertweck馆长说，这样的架空建筑"释放了地面空间，这一空间无论是从建筑本身来说，还是从象征意义上来说，都是自由的"，"虽然我们作为建筑师不能改变政治，但我们可以通过建筑工程和建筑类型，也通过与客户协商，为公众提供更多空间。"

2018年在黎巴嫩举办的贝鲁特设计周的主题是"设计和城市：__________"，其关注点是公共建筑领域的代理机构和代表。贝鲁特也越来越多地受到私有化的影响。就在去年，黎巴嫩最后一个公共海滩——拉姆莱特·艾尔·拜达的相当大一部分被一个豪华度假村所取代。[8]但是，本次活动展出的作品，包括弹出式图书馆、可移动长凳、香烟烟头容器和空中花园等，都鼓励草根阶层采取行动来改变户外空间。

让我们从公共空间转向私有空间。社会学家亨利·列斐伏尔认为，城市是日常生活节奏通过感官体验得以展现的场所。但是感官上的城市空间有被均质化侵蚀的危险。克里斯托弗·德沃夫在他的论文《香港的感官导航》[9]中描述道："香港的城市形态何其复杂，只有身在其中、深感其内的人才能感受到——且是切身直观地感受。"当游客从潮湿的户外通过桥梁、道路和通道走到有空调的购物中心时，他们会受到各

the bizarre. Loitering, standing in groups, busking, selling lucky charms, and even bringing sheep into a village have been banned in some localities.[5]

It is not hard to view with cynicism the commercialism of many modern plazas, designed as they are, to encourage foot-fall into adjacent cafés and shops. Their calming greenery ensures that lunching workers (or should I say "hands"), having parted with their money for coffee, will return to their offices in the afternoon with recharged, more productive brains.

We also rue their sanitised condition, which limits the unexpected and excludes the undesirable. As Helle Juul writes in *Public Architecture: The Familiar into the Strange*, a city's strength is its diversity, where people of diverging interests and affiliations rub along despite differences. When common space is "driven away by private interests and self-serving considerations, the society's cohesive force vanishes and concomitantly its potential to evolve."[6] If every thing or body that makes a mess or disturbance is unwelcome, this is frightening indeed.

But architecture can also come to the rescue. In Luxembourg, just 8% of land is owned by the state.[7] At this year's Venice Biennale (on the theme of FREESPACE – a telling choice of subject-matter) Luxembourg pavilion's exhibition "The Architecture of the Common Ground", displays large MDF models of architects' unrealised projects which maximise space beneath raised buildings. Le Corbusier's unbuilt Îlot insalubre n°6, would have included a kindergarten, cinema and library, for instance. Such elevated buildings "leave the ground that they stand on physically and symbolically free," say curators Andrea Rumpf and Florian Hertweck. "Even if we can't change politics as architects, we can react with programmes, with typologies and maybe negotiate with the client to give more space to the public."

In Lebanon, Beirut Design Week 2018's theme, "Design and the City: __________", specifically looked at agency and representation in the public realm. Beirut is increasingly impacted by privatisation too – only last year a sizeable chunk of the

蓝色住宅，香港
The Blue House, Hong Kong

种噪声、异味和高温的攻击。这些体验有助于在自身记忆体系中建立个人感官地图，帮助自己识途辨路。但是，在这个快速发展的快节奏城市，发展往往“忽略了城市密集的人口和活动所赋予的丰富性”，并且“对香港一贯特有的一幢建筑多种用途的做法零容忍”。

相比之下，德沃夫则将蓝色住宅称为一个包容其混乱历史遗产的项目。蓝色住宅是建于20世纪20年代的一座公寓大楼，曾经是一家医院、一所功夫会馆、一座寺庙和一个骨科诊所的所在地。政府已经决定对此开发重建，但是社会工作者和自然资源保护主义者说服政府保留了这里的随意风格以及其随性的蓝色。公寓翻新了，增加了浴室和其他基本设施，但保留了其原有特色，如仍保留了原建筑内的瓷砖、楼梯和五金制品。现在，在公寓大楼的中心位置有一个社区活动空间，里面还有两家餐厅和一个展示该公寓历史风貌的博物馆式空间。大多数原有居民都选择继续住在这里，所有新来的居民也都完全接受这一社区精神。教科文组织曾赞扬这一做法：“在世界上压力最大的房地产市场之一的地方，这种前所未有的保护边缘化地方遗产的公民努力，对保护该地区乃至其他城市地区遭受四面楚歌的古老建筑来说都是一种极大的鼓舞。”[10]

在建成环境中给人良好个人感官体验的另一个建筑实例是公共游泳池。由BIG建筑事务所设计的奥尔胡斯海港浴场，沿着滨水区有一系列休闲娱乐设施，市民可以在海陆之间的狭长地带进行锻炼和玩耍，这里有悬崖峭壁、浮桥、码头和干船坞等后工业时代的台地景观。

游泳池和浴池可以让人们从身体上摆脱身份的束缚；人们在水上漂浮时感受到的自由自在就跟被人们认可并接受的裸体所赋予人们的自由一样。这些地方的建筑给人独特的嗅觉和触觉感受：溅起的水珠、浸泡的感觉、氯气的气味、令人目眩的光线的反射和折射、朦胧的SPA中心里的朦胧形态，以及公共更衣室和淋浴间更模糊的边界。许多文化中都有个人仪容仪表与社交活动相结合的传统，其中包括罗马的温泉浴场、（伊朗的）公共澡堂、（日本的）温泉或（韩国的）休闲养生水疗中心。这些地方都使私人和公共以及室内和室外的概念变得模糊不清；这

last public beach – Ramlet al Baida – was replaced by a luxury resort.[8] But exhibits including a pop-up library, a moveable bench, a cigarette waste receptacle and a garden on stilts all encouraged grassroots action to transform outdoor spaces.
From the pseudo-public, let's turn to the pseudo-private. Sociologist Henri Lefebvre argued that the city is the setting where the rhythm of everyday life is played out in sensory experiences. But sensuous urban space risks erosion by homogenisation. In his essay "Sensory Navigation in Hong Kong"[9], Christopher DeWolf describes the city as having "a complexity of urban form which is legible only to those who have developed a sense for it – quite literally so." A visitor is assaulted by different noises, smells and temperatures as they walk from the humid outdoors to air-conditioned malls via bridges, roads and passageways: these experiences help build a personal sensory map in people's memories which allow them to navigate their way around. But in the fast-paced and growing city, developments often "overlook the richness afforded by the city's density of people and activities" and have "no tolerance for the mish-mash of uses that has always characterised Hong Kong".
In contrast, DeWolf cites the Blue House as a project which has embraced its chaotic heritage. The 1920s tenement block, once home to a hospital, a kung fu master, a temple and an osteopathy clinic, had been earmarked for development. But social workers and conservationists persuaded the government to preserve the building's haphazard character (and arbitrary blue colour). The flats were modernised with bathrooms and other essential amenities but retained various eccentricities, such as tiles, staircases and ironmongery. There is now a community space in the center, two social-enterprise restaurants and a museum-type space showcasing the apartments' historic appearance. Most of the existing residents have remained and any newcomers fully embrace the community ethos. UNESCO praised the development: "This unprecedented civic effort to protect marginalised local heritage in one of the world's most high-pressure real estate

BIG设计的奥尔胡斯海港浴场
Aarhus Harbor Bath by BIG

JKMM建筑师事务所设计的阿莫斯·瑞克斯博物馆
Amos Rex Museum by JKMM Architects

些地方没有过多条条框框的限制，扮演社会平等者的角色，同时发挥其文化和实用的功能。

现在回到私有公共空间这一概念。在赫尔辛基的拉西帕拉特西广场下面，有一座由JKMM建筑师事务所设计的阿莫斯·瑞克斯博物馆。它是阿莫斯·安德森艺术博物馆的新家，后者是一家私人拥有的画廊，以其捐建人（一家当地报纸出版商）的名字而命名，于20世纪60年代在其去世后成立。地下画廊似乎是通过自己的天花板向上拱出地面进入了地表的公共领域，呈现出一个锥形土丘的月球景观。这些沙丘状天窗对参观博物馆的游客和滑板爱好者来说是一个有趣的、富有触感的环境。

有趣的是，该博物馆归私人所有的属性提高了其知名度。经过五年的斡旋努力后，在赫尔辛基建一座古根海姆博物馆的提议还是遭到否决。因为它是作为公私合作的建筑项目被构想出来的，人们否决这一提议的一个重要因素是担心会因为出资修建这一项目而导致其他公共项目经费遭到削减。[11] 阿莫斯·瑞克斯博物馆是一个极具原创性的文化中心和开放空间，提供了独特的感官体验，展览展示了芬兰人的才华。

在财政紧缩的时代，资金短缺的市政当局无法为公共设施的更新甚至基本的维护提供资金，为我们的城市提供所需的公共设施的责任就落在了商业土地所有者和开发商的肩上。虽然无限制的私有化确有问题，但大声疾呼的二元论点也是有问题的。二元论观点非常滑稽地将公共和私有从本质上区分好与坏，却没有将它们视为存在于一个有机体中具有细微差别的实体。

本书的建筑实例展示了当利益相关者合作时，如何成功地处理好“公共”和“私有”的关系。连续一致的政府法规、对建筑有效的维护维修、企业的社会责任和透明的沟通是考虑的最基本因素。同时从事建筑行业的设计人员需要具有独特的设计能力并提供敏感而富有想象力的解决方案。

markets is an inspiration for other embattled urban districts in the region and beyond."[10]

Another example of personal sensory experience within the built environment is the public pool. At Aarhus' Harbour Baths, designed by Bjarke Ingels Group (BIG), a series of recreational leisure facilities along the waterfront allow citizens to exercise and play in a liminal zone between land and sea, in a terraced post-industrial landscape of cliffs, pontoons, piers and dry docks.

Swimming pools and baths allow people to physically divest themselves of the trappings of status; the freedom of floating is mirrored by the freedom granted by sanctioned nudity. The architecture of such places is imbued with unique olfactory and tactile sensations: splashing water, immersion, chorine, disorienting reflections and refraction of light, hazy forms in a misty spa, and the further blurring of boundaries in communal changing rooms and showers. Many cultures have traditions where personal grooming is mixed with socialising, including the Roman thermae, the hamam, the Onsen or the Jjimjilbang. These all make the concepts of private and public and indoors and outdoors indistinct; they loosen inhibitions, act as social levellers, and fulfil both a cultural and practical function.

Back to POPS. In Helsinki, beneath Lasipalatsi Square, sits JKMM Architects' Amos Rex Museum. This is a new home for the Amos Anderson Art Museum, a privately-owned gallery founded after the death in the 1960s of its eponymous patron, a local newspaper publisher. The subterranean gallery appears to bubble up through its own ceiling into the public realm above, creating a moonscape of cone-shaped mounds. These dune-like skylights are a fun, tactile environment for museum-goers and skateboarders alike.

Interestingly, the museum's private ownership has improved its popularity. A proposed Guggenheim Museum for Hel-

精心设计的公共建筑的好处众多，这一点是不言而喻的。这些建筑和空间对历史、环境非常敏感，具有对其周围环境独特的感官营造品质，如果（也许这有些天真？）这些建筑和空间得到充分管理和经营，就能够创造出令人愉快的、人人可以享用的环境。这些建筑和空间可以培养公民的自豪感，增强社区意识，提高个人幸福感。最好，这些建筑和空间不限制公民权利，没有歧视，没有边缘化，而是促进凝聚力，消除不平等。从经济方面来说，这些颇具吸引力的场所能支持现有的商业企业，增加信心，鼓励创业。以上这些要求听起来是不是天方夜谭不可能实现？尽管如此，我们一样会要求建筑师这样做。

1. Ian Nairn, 'no two the same', Cinemagazine, 1970, BFI player
2. Bansky has his own interpretation of what constitutes privacy and property in the built environment. Quote taken from interview with Timeout Magazine, 'Street (il)legal', David Fear, April 12 2010
3. 'Revealed, the Insidious Creep of Pseudo-Public Space in London', The Guardian, July 24 2017
4. Singapore Urban Redevelopment Authority, 'Good Practice Guide for POPS', https://www.ura.gov.sg/Corporate/Guidelines/Circulars/dc17-02
5. 'Councils are using these new rules to ban pretty much everything', Josie Appleton, Architecture Foundation, 4 August 2016. https://www.architecturefoundation.org.uk/writing/public-freedom
6. Juul | Frost Architects, Public Architecture: The Familiar into the Strange, Copenhagen, 2011
7. LUCA Luxembourg Centre for Architecture, architecturebiennale.lu
8. Beirut's last public beach: residents fear privatisation', The Guardian, 2 February 2017
9. 'Sensory Navigation in Hong Kong', Christopher DeWolf, 2017 (in The Future of Public Space, SOM Thinkers, New York, 2017 pp 80-90).
10. 'Hong Kong's historic Blue House wins Unesco's highest heritage conservation award', Naomi Ng, South China Morning Post, 01 November 2017
11. 'Amos Rex is Helsinki's Homegrown Star', Nina Siegal, New York Times, September 5 2018

sinki was rejected after a 5-year battle, a major deciding factor being the concern that – as it was conceived as a public-private partnership – other public projects would suffer cutbacks to fund it.[11] Amos Rex is a highly original cultural center and open space which provides a unique sensory experience and showcases Finnish talent.

In an age of austerity, where cash-strapped municipalities are unable to finance the regeneration or even basic maintenance of public amenities, it falls to commercial landowners and developers to provide the facilities which our cities need. Whilst unrestrained privatisation is problematic, so are vociferous, binary arguments which caricature public vs private as intrinsically good vs bad without recognising them as nuanced entities which exist on a spectrum.

The examples in this book demonstrate how "public" and "private" can be handled successfully when stakeholders collaborate. Coherent government regulations, effective maintenance, corporate social responsibility and transparent communication are fundamental considerations. Members of the architecture profession have the unique design abilities to provide sensitive and imaginative solutions.

The benefits of well-designed public architecture are manifold and self-evident. Assuming (perhaps naively?) that they are managed adequately, such buildings and spaces, sensitive to history, context and the unique sensory place-making qualities of their surrounds, create enjoyable, accessible environments. They foster civic pride, boost a sense of community and enhance individual well-being. At their best, they are nurturing places which do not restrict the rights of citizens, discriminate or marginalise, but rather promote cohesion and level inequalities. Economically, these attractive places support existing businesses, increase confidence and encourage entrepreneurship. An impossible list of requirements? We will demand it of our architects all the same.

帝国商店
Empire Stores

S9 Architecture + STUDIO V Architecture

帝国商店是布鲁克林从一个衰败的工业巨头转型为一个不断增长的创意产业区的象征。本案这个屡获殊荣的综合用途开发项目重新设计了位于DUMBO海滨的建于19世纪的一个闲置大型砖砌仓库。今天，DUMBO是布鲁克林蓬勃发展的科技铁三角中三个关键区域之一，也是超过500家科技公司及近10 000个技术工作岗位的创新孵化器，并且其规模还在不断增长。这个41 806m^2的改扩建综合体建筑群，以及一栋两层共557m^2的屋顶加建建筑，提供了急需的办公空间，并将零售、餐饮、公共空间和艺术展览馆引入了该区域。这一适应性再利用项目彰显并保护了这座海滨建筑的不朽存在，同时改善了DUMBO城市网和34.4ha布鲁克林大桥公园之间的交通。

多年来，这座堡垒般的建筑一直高高耸立，阻挡了公园与繁华的城市区域之间的视野。S9建筑设计事务所把握住了机会，将帝国商店建筑群稍加改动，用作连接两个步行区的门户。从砖石结构中切割出来的通道在水街和海滨之间形成了一条人行通道。从该建筑物中心开凿出的四层露天庭院为租户、社区成员和公园游客提供了一个沉浸式公共空间。公园与庭院之间的通道自由穿过一层没有安装玻璃的巨大拱形窗户。庭院内的玻璃幕墙融现代风格和历史风格于一体，使建筑的层次和功能一目了然：位于一层的是购物广场、公共美食广场；布鲁克林历史学会的画廊在二层；上面几个楼层是开放式办公空间。一部宏伟而不凡的皮拉内西风格楼梯一直延伸到二层的夹层，楼梯兼作圆形剧场式座位，用作在庭院举行公共活动和观看表演时的观众座席。

建筑师通过将屋顶改造成可从庭院进入的公共景观露台，使布鲁克林大桥公园与该建筑有机地结合在一起。露台花园采用节水景观，并采用拆除过程中回收的材料制作定制座椅和固定装置。壮观的屋顶露台扩展了人们户外休闲和交际活动的公共空间。这里增添了公园的娱乐设施，设有餐厅和啤酒花园。此外，在这里，人们亦可欣赏到大桥和曼哈顿天际线的标志性景观。现在，该建筑把DUMBO居民区与公园的壮丽景色、公共艺术/文化活动和广泛的娱乐设施结合起来。

这个经过翻修而充满活力的建筑群拥有五层共35 303m^2的创意办公空间，其中包括屋顶上的两层当代加建建筑。零售店和餐厅位于一层，面积为6503m^2，二层有278.7m^2的展览空间。

帝国商店使这座历史悠久的建筑焕然一新，它的设计满足当代需求，从整体上考虑终端用户的需求，同时尊重原有的建筑和场地。帝国商店再次成为布鲁克林海滨的一颗璀璨的宝石，且在布鲁克林的历史性跨越成长中发挥了重要作用，并将继续作为活动中心连接商业、文化和社区。

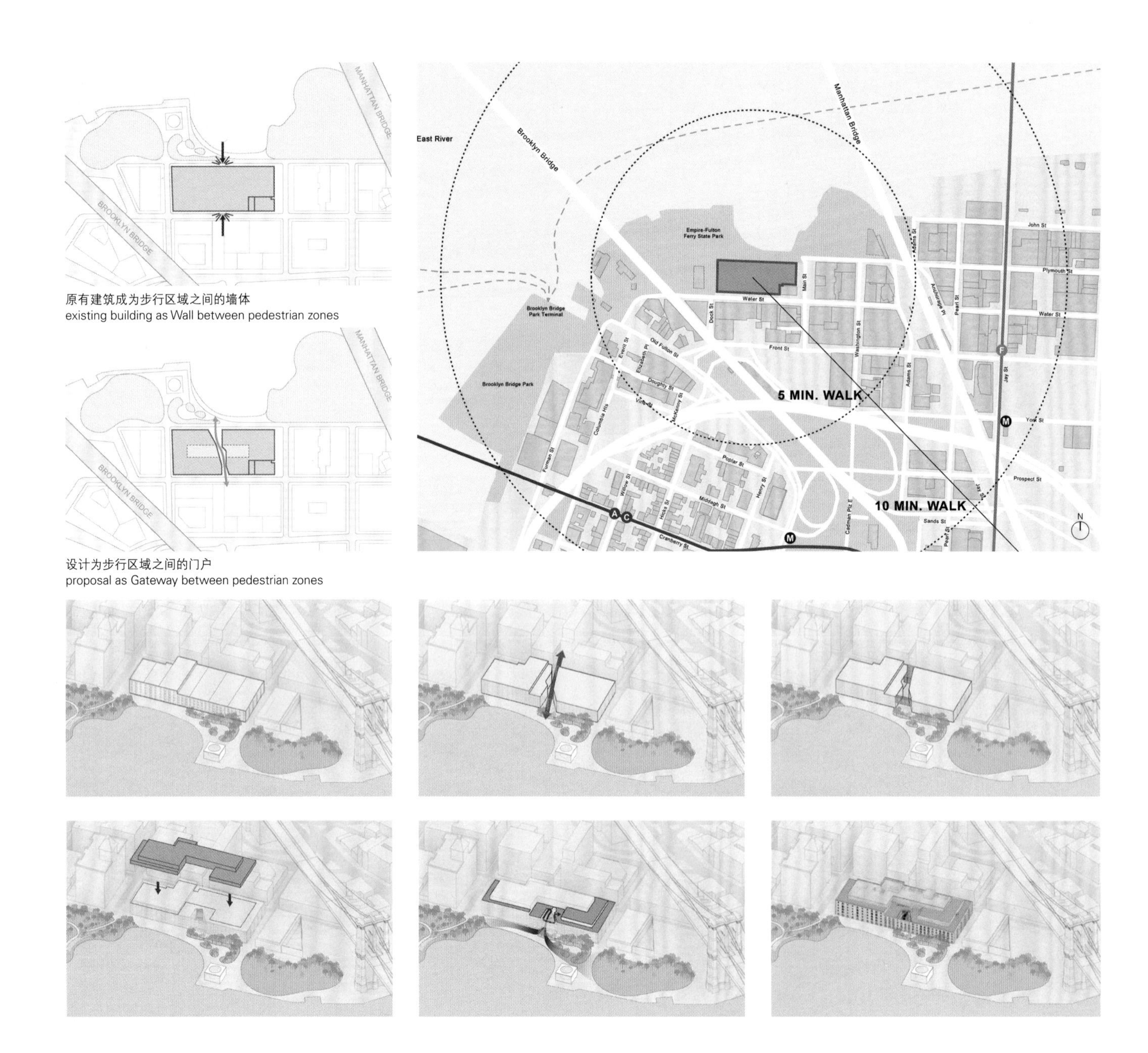

原有建筑成为步行区域之间的墙体
existing building as Wall between pedestrian zones

设计为步行区域之间的门户
proposal as Gateway between pedestrian zones

Empire Stores is emblematic of Brooklyn's transformation from a lapsed industrial powerhouse into a growing creative sector. This award-winning mixed-use development reimagines a vacant, 19th century massive brick warehouse on the DUMBO waterfront. Today, DUMBO is one of the three key zones in Brooklyn's burgeoning Tech Triangle, an innovation incubator home to over 500 tech firms, and nearly 10,000 tech jobs, and counting. The conversion and expansion of this 41,806m² complex, as well as a two-storey, 557m² rooftop addition, provides much-needed office space and brings retail, dining, public space and exhibition galleries to the neighborhood. The campaign of adaptive re-use celebrates and preserves the building's monumental presence on the waterfront while improving circulation between DUMBO's urban fabric and the 34.4 ha Brooklyn Bridge Park. For years, the fortress-like building stood as a visual and physical barrier between the park and the thriving urban neighborhood. S9 Architecture saw an opportunity to repurpose the Empire Stores complex as a gateway portal linking the two pedestrian zones. A passageway carved out of the masonry structure creates a pedestrian conduit between Water Street and the waterfront. A four-storey, open-air courtyard excavated from the center of the building serves as an immersive public space for building tenants, community members, and park visitors. Access between the park and the courtyard is free flowing through the unglazed, monumental arched windows at ground level. Glass curtain walls lining the courtyard blend the contemporary and the historic to make visible the building's striations and function: shopping and a public food court at courtyard level, galleries for the Brooklyn Historical Society on the first floor, and multiple floors of open office

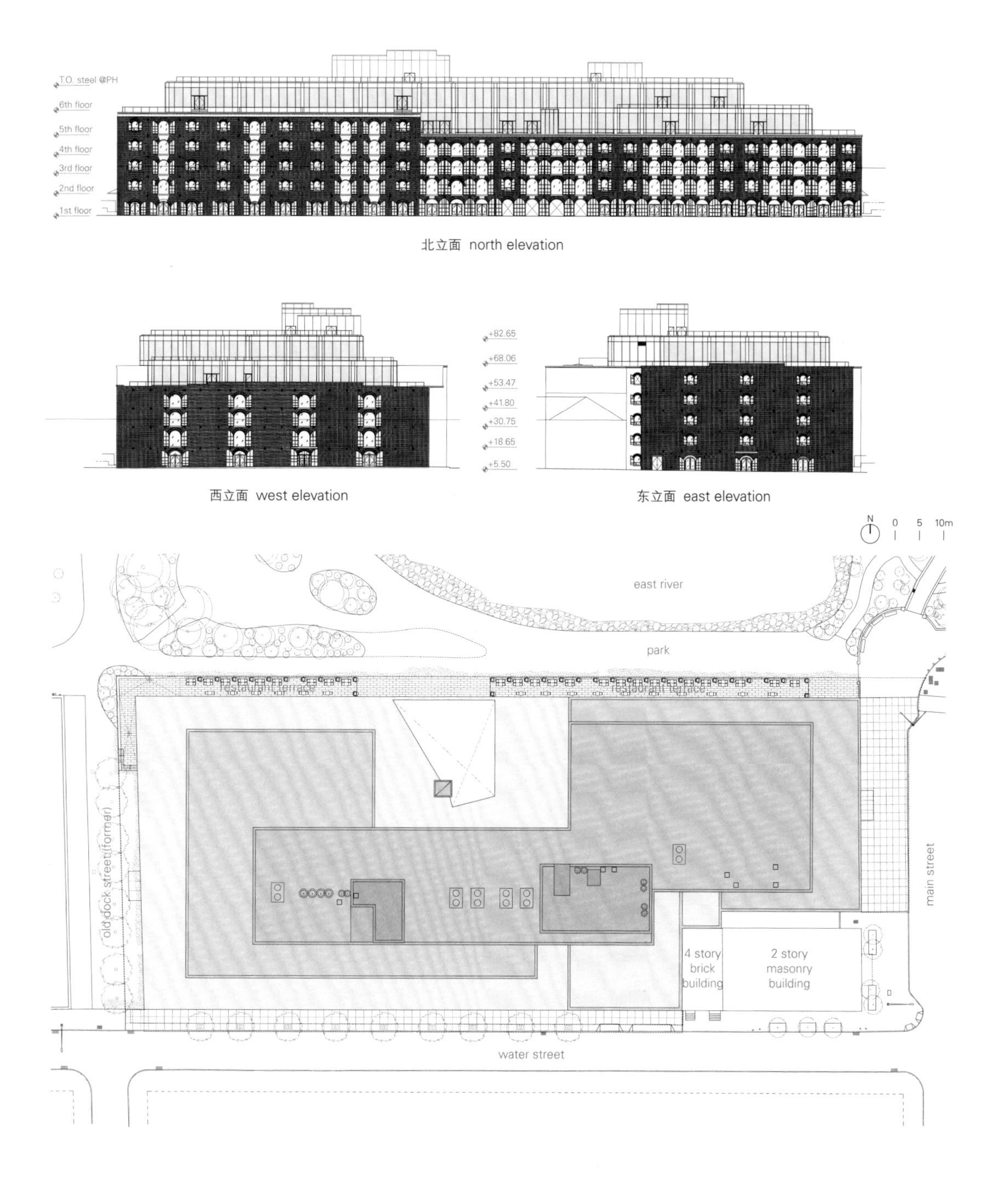

北立面 north elevation

西立面 west elevation

东立面 east elevation

VHH
CECCONI'S
SMILE
TO GO
FEED
BROOKLYN
HISTORICAL
SOCIETY
SHINOLA
DETROIT
PUBLIC
RESTROOM
west elm
SUGARCANE
EMPIRE

EMPIRE
STORES
CECCONI'S
FEED
SHINOLA
west elm
A1

DUMBO

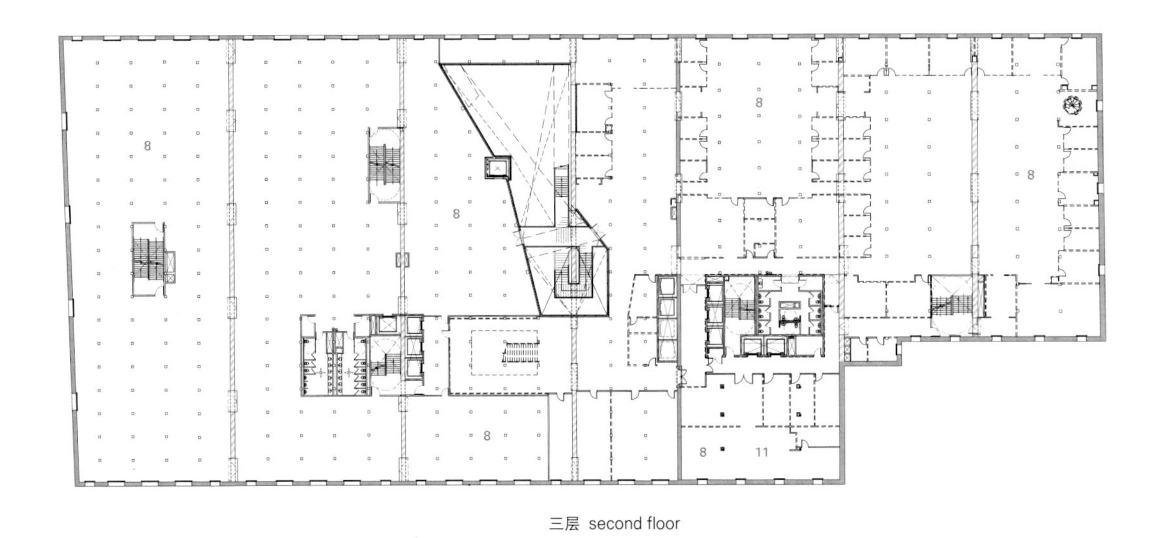

三层 second floor

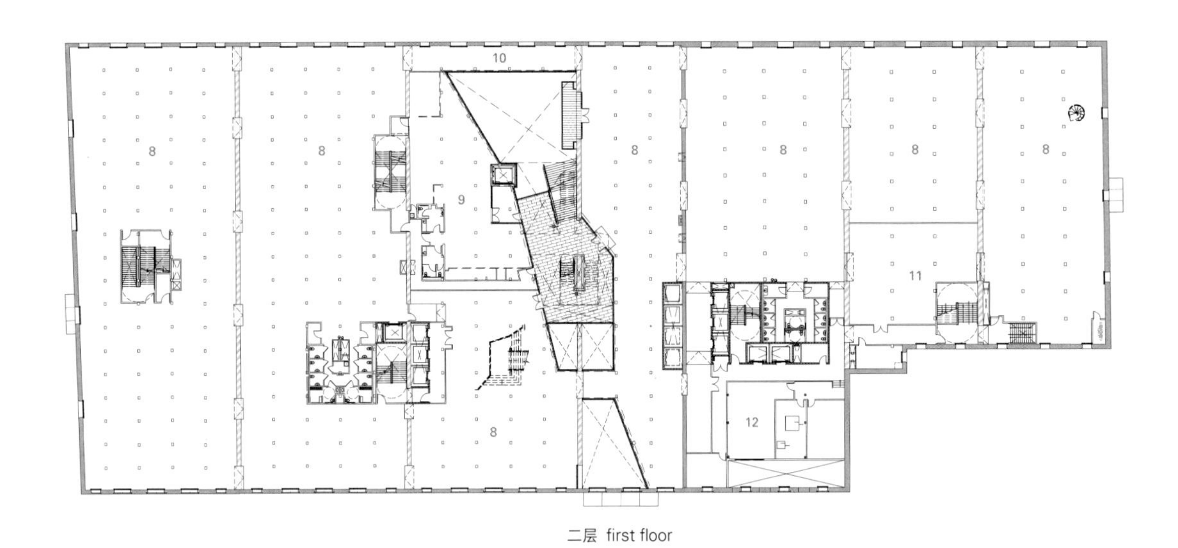

二层 first floor

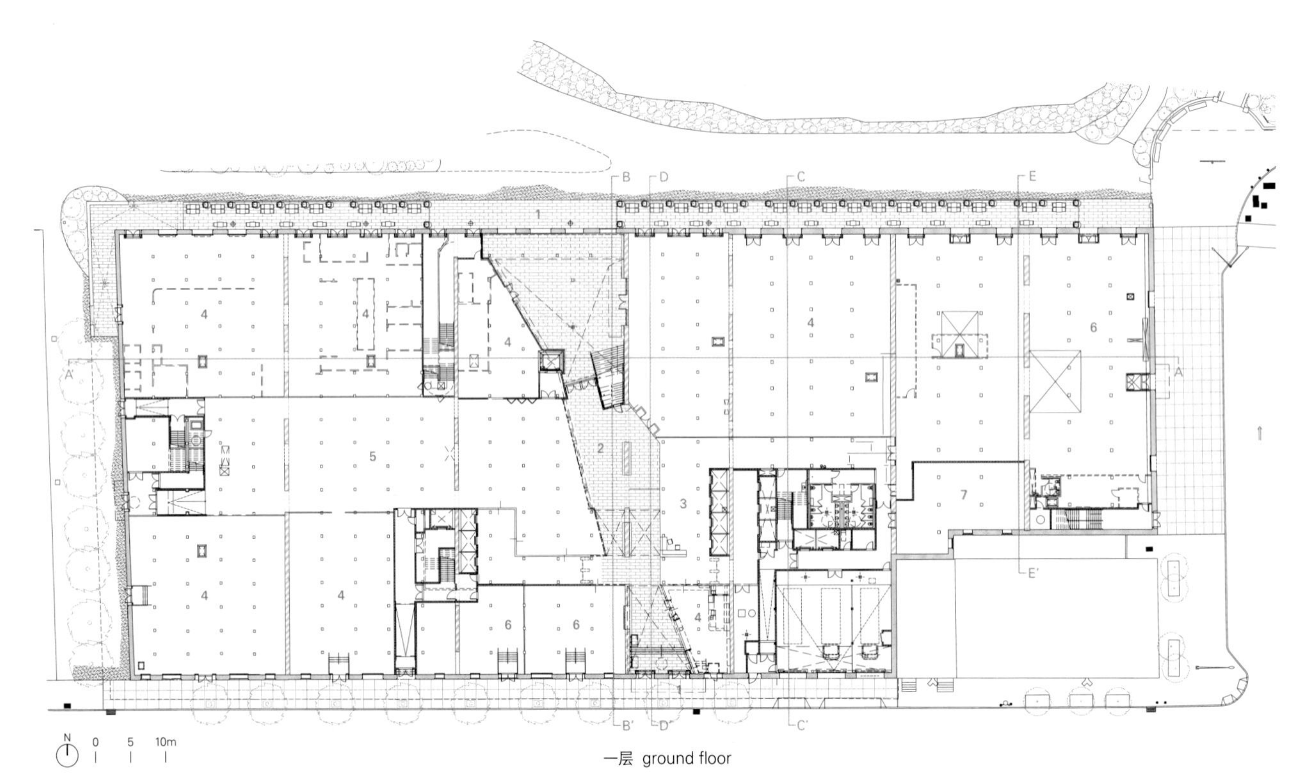

一层 ground floor

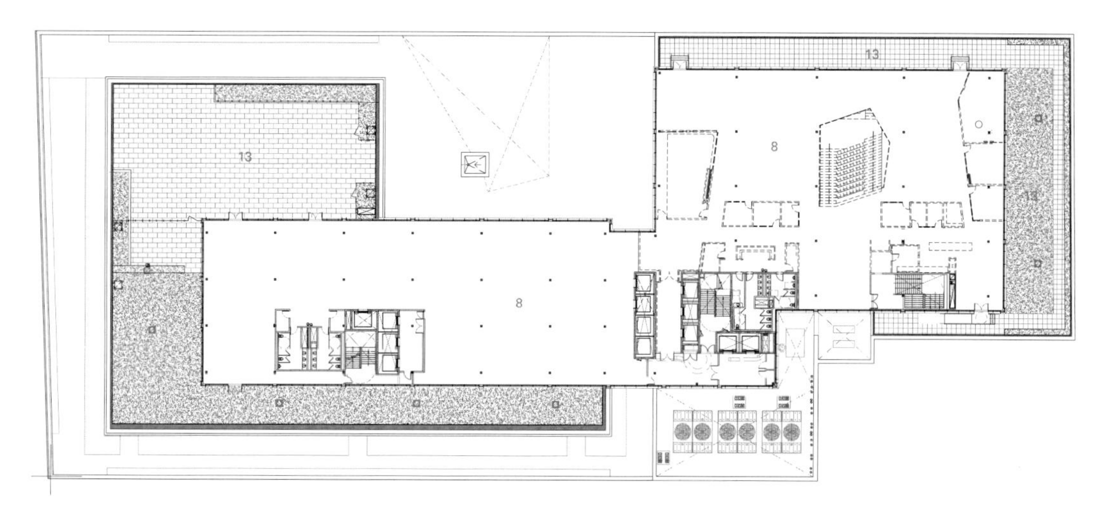

六层 fifth floor

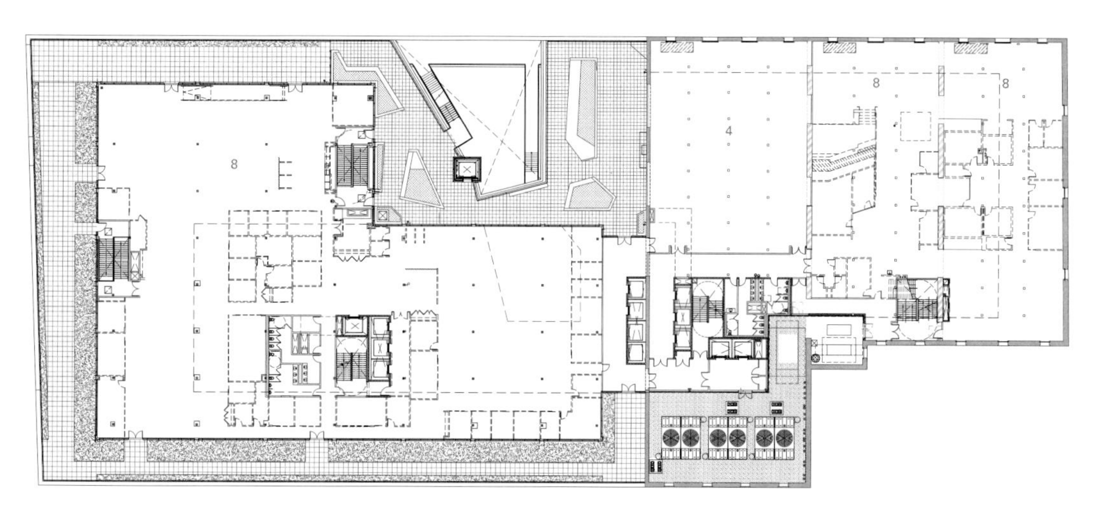

五层 fourth floor

1. 入口 2. 庭院 3. 大厅 4. 餐厅 5. 市场 6. 零售店 7. 零售店储藏室 8. 办公室 9. 博物馆 10. 步道 11. 储藏室 12. 机械室 13. 露台
1. entrance 2. courtyard 3. lobby 4. restaurant 5. market 6. retail 7. retail storage 8. office 9. museum 10. walkway 11. storage 12. mechanical 13. terrace

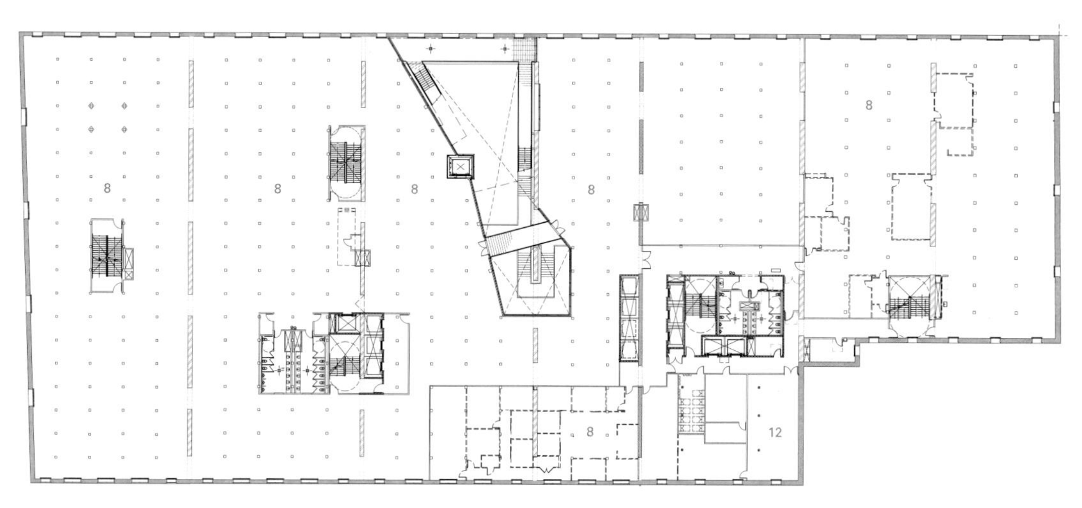

四层 third floor

项目名称：Empire Stores / 地点：55 Water Street, Brooklyn, New York / 事务所：S9 Architecture, Studio V Architecture / 结构工程师：Robert Silman Associates / 机电管道工程师：Mottola Rini Engineers / LEED顾问：Spiezle Architectural Group / 照明顾问：Tillotson Design Associate / 景观设计师：Future Green Studio
总承包商：Veracity Partners / 客户：Midtown Equities, HK Organization and Rockwood Capital / 用途：retail, restaurant, office, mixed use / 建筑面积：41,800m² (450,000 SF) / 竣工时间：2017 / 摄影师：©Imagen Subliminal (courtesy of the S9 Architecture) (except as noted)

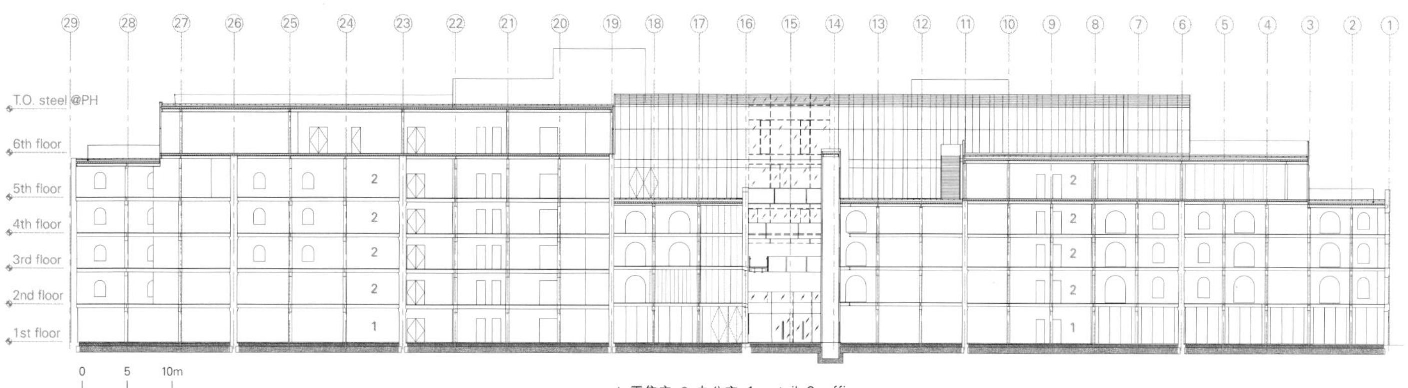

1. 零售店 2. 办公室 1. retail 2. office
A-A' 剖面图 section A-A'

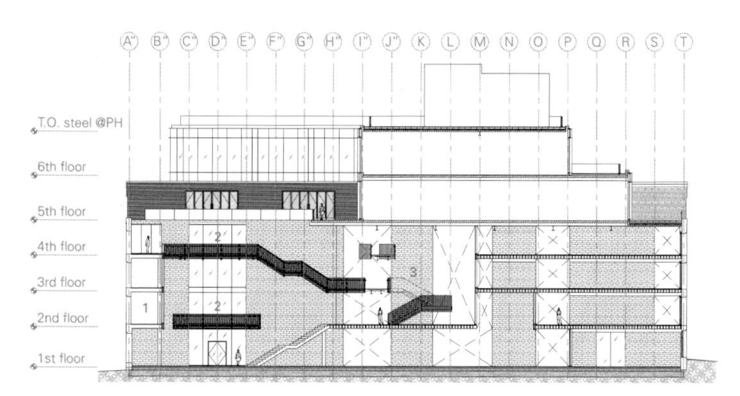

1. 步道 2. 露台 3. 庭院 1. walkway 2. terrace 3. courtyard
B-B' 剖面图 section B-B'

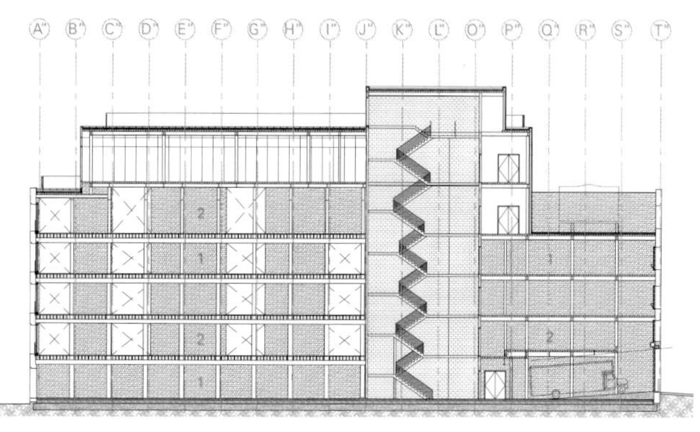

1. 餐厅 2. 办公室 1. restaurant 2. office
C-C' 剖面图 section C-C'

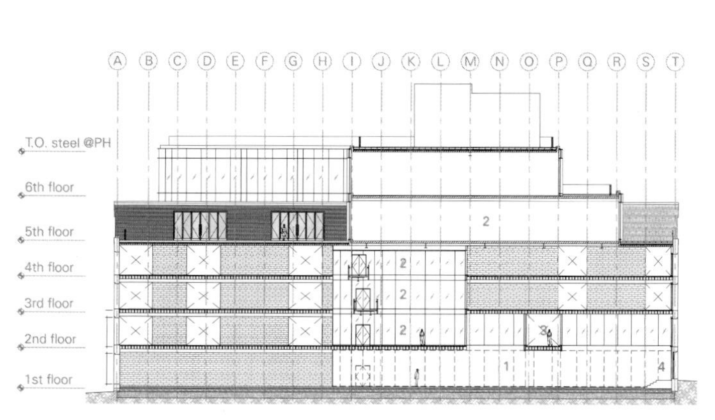

1. 餐厅 2. 办公室 3. 连接桥 1. restaurant 2. office 3. bridge 4. entrance
D-D' 剖面图 section D-D'

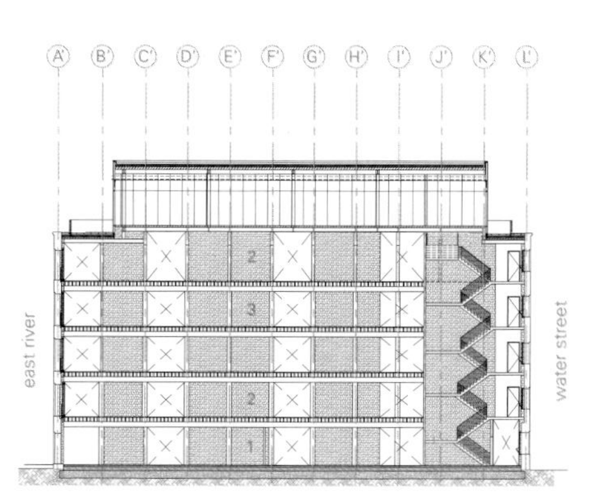

1. 零售店 2. 办公室 3. 餐厅 1. retail 2. office 3. restaurant
E-E' 剖面图 section E-E'

space above. A grand, Piranesi-inspired staircase ascends to a first-floor mezzanine and doubles as amphitheater-style seating for public events and performances taking place in the courtyard.

By adapting the rooftop into a landscaped public terrace accessible from the courtyard, Brooklyn Bridge Park extends into the building organically. The terrace gardens utilize water-efficient landscaping and incorporate custom seating and fixtures produced from materials reclaimed during demolition. This spectacular rooftop terrace provides expanded communal space for outdoor leisure and interaction. This space augments the park's recreational facilities with a restaurant and beer garden and offers iconic views of the bridges and the Manhattan skyline. Now, the building facilitates access between the residential enclave of DUMBO and the park's breathtaking views, public art/cultural events,

and extensive recreational facilities.

The reanimated complex features 35,303m² of creative office space over five floors, including a two-storey contemporary addition on the roof. Retail and restaurants constitute 6503m² on the ground floor with 278.7m² of exhibition space on the first floor.

Empire Stores brings new life to a historic building, re-envisioned for contemporary needs, holistically considering the needs of the end user, while respecting the context of the existing building and site. Once again, Empire Stores resumes pride of place as a jewel of the Brooklyn waterfront, a structure that plays a significant role in Brooklyn's historic growth, and will continue to do so as a contemporary hub linking business, culture, and community.

THEATER
SHINOLA

©Patrick Donahue (courtesy of the S9 Architecture)

©Patrick Donahue (courtesy of the S9 Architecture)

©Patrick Donahue (courtesy of the S9 Architecture)

©Patrick Donahue (courtesy of the S9 Architecture)

大型商场步道
Mega Foodwalk

FOS Foundry of Space

“通过流畅的购物体验使城市生活重新融入自然”

Megabangna购物中心的规模堪比一座小镇。它的中央建筑作为商业中心区而存在，而东翼的美食街则被规划成有着更多绿地和运河的乡村景观。零售区的扩建结构位于原有区域以外的东侧边缘，概念上它是一片“山谷”，这是最宜人的自然风貌之一，其中也有连绵葱郁的群山环绕的私密中央空间。

新扩建结构的建筑概念是“山谷”，源于其隐喻的地理特征。为了营造与天然山谷相似的氛围，新的开放式购物中心的布局围绕着一个中央庭院空间展开，其中一个下方带有圆形剧场的下沉广场作为顾客的主要社交空间，用于集会和举行各种活动。

从底层的下沉广场继续走，就会来到布局中间被称为“山丘”的倾斜绿色区域，这里平缓地延展上升，与一层现有的大型广场平稳连接。

“山丘”旨在打造一个休闲空间，人们可以完全沉浸在郁郁葱葱的景观中，还能欣赏水景和体验户外设施。建筑师在开放式庭院和整座建筑物中加入了绿色植物，从而使该建筑成为一个市场和公共公园的混合体，人们更愿意在这里与他人开展社交活动。

同时，通过连廊和室内步道，每个楼层店面外的步道网络与原有的步道走廊和新停车场连接了起来。这样，在两个大型建筑群之间就形成了没有死胡同的无缝连接交通流线。

此外，将自然环境转变为独特的购物体验这一想法，通过合理组织空间和结合各种建筑元素得以实现。这里连续设置了一系列的倾斜走道，它们的倾斜度最小角度为1:15，从上到下缓缓下降，给人们带来一种“山地步行”般的体验。这不仅有效地增加了低楼层销售区域的面积，而且还形成了无限螺旋循环的步道，盘旋连接着四个楼层。

"Reconnecting Urban Life with Nature Through a Flowing Shopping Experience"

The sheer size of Megabangna shopping complex is as large as a small town. Its central building is perceived as a downtown, whereas Foodwalk zone on the east wing is portrayed as countryside with more green areas and canals. The new extension of retail zone located on the eastern periphery beyond the existing zone could then be conceptualized as a "Valley", one of the most pleasant natural topography in which its intimate central space is enclosed by continuous frontage of lushly mountains.

The architectural concept of the new extension, "The Valley", therefore derives from the geographic character of its metaphor. To create similar atmosphere to a natural valley, the layout of the new open-air mall is composed around a central courtyard space, in which a sunken plaza with an amphitheatre down below acts as a customers' main social space for gathering and holding all kinds of events. Continuing from the sunken plaza on the bottom level, the sloping green area in the middle of the layout, called "The Hill", gently ascends to connect smoothly with the existing Mega Plaza on level 1. The Hill is intended to be a relaxing

space where people can fully immerse themselves into the lush landscape with water features and outdoor furniture. By embedding a lush greenery into the open-air courtyard and throughout the building, the project becomes a hybrid of a market place and a public park where social interactions are more encouraged among people.
Simultaneously, network of walkways along shopfronts on every level is connected to the existing corridor and a new car parking building via link bridges and covered walkway in order to complete a seamless circulation system between the two phases without dead end.

Moreover, the idea of transforming natural environment into a unique shopping experience is synthesized through its spatial organization and various architectural elements. A series of minimum 1:15 sloping walkways are positioned continuously, descending gently down from upper to lower levels, to create a similar experience of "hill walk". It effectively results in not only increased saleable areas on the lower floor but also an infinite loop of spiral circulation, circling endlessly on all four levels.

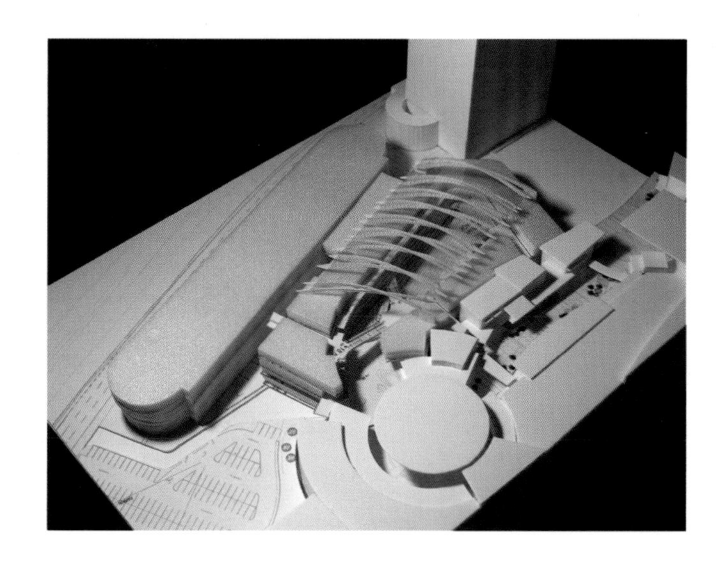

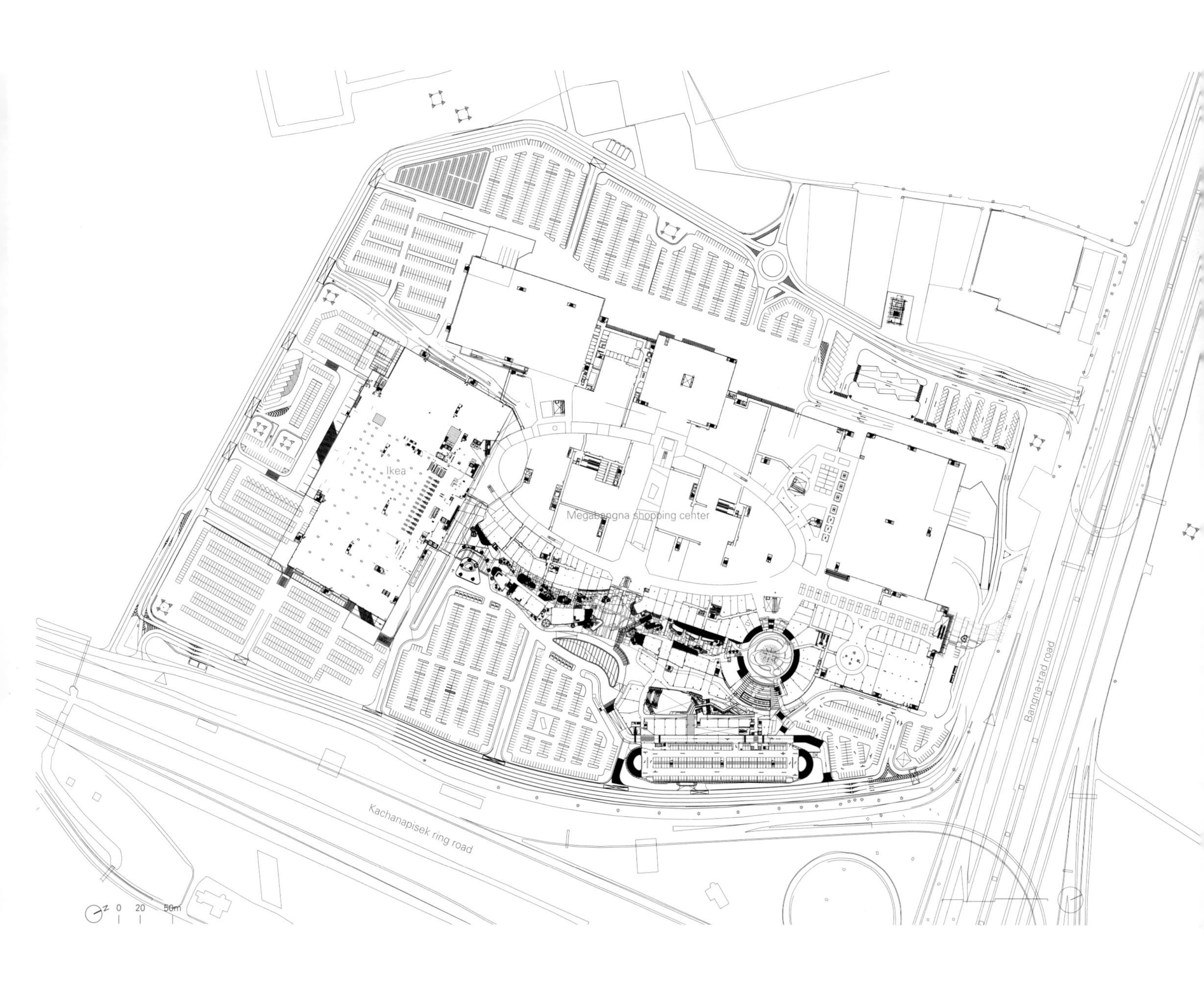
Ikea
Megabangna shopping center
Kachanapisek ring road
Bangna-trad road
0 20 50m

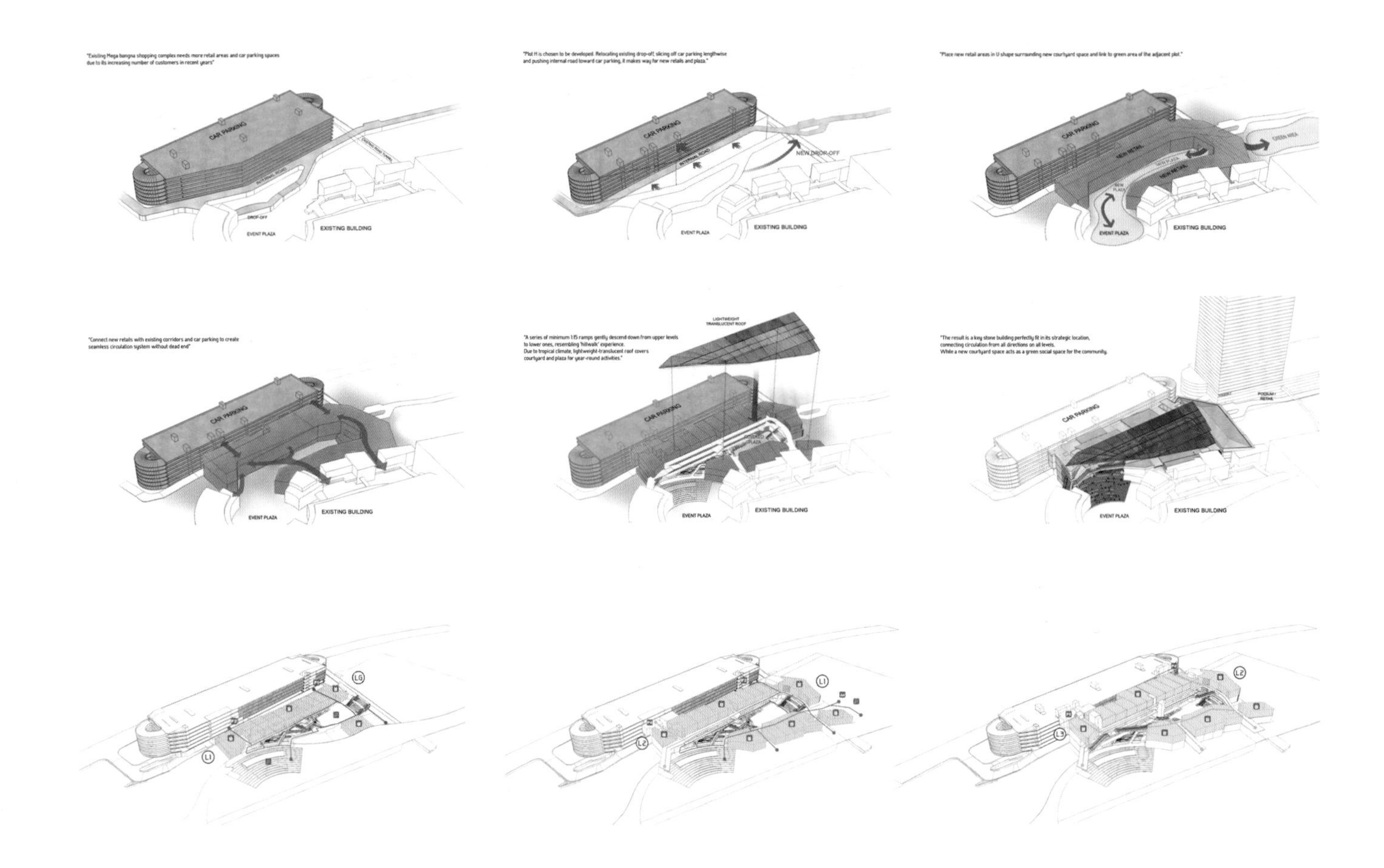
"Existing Mega bangna shopping complex needs more retail areas and car parking spaces due to its increasing number of customers in recent years"
CAR PARKING
EVENT PLAZA
EXISTING BUILDING
"Plot H is chosen to be developed. Relocating existing drop-off, slicing off car parking lengthwise and pushing internal road toward car parking, it makes way for new retails and plaza."
CAR PARKING
INTERNAL ROAD
NEW DROP-OFF
EVENT PLAZA
EXISTING BUILDING
"Place new retail areas in U shape surrounding new courtyard space and link to green area of the adjacent plot."
CAR PARKING
NEW RETAIL
GREEN AREA
EVENT PLAZA
EXISTING BUILDING
"Connect new retails with existing corridors and car parking to create seamless circulation system without dead end"
CAR PARKING
EVENT PLAZA
EXISTING BUILDING
LIGHTWEIGHT TRANSLUCENT ROOF
"A series of minimum 1:15 ramps gently descend down from upper levels to lower ones, resembling 'hillwalk' experience.
Due to tropical climate, lightweight-translucent roof covers courtyard and plaza for year-round activities."
CAR PARKING
EVENT PLAZA
EXISTING BUILDING
"The result is a key stone building perfectly fit in its strategic location, connecting circulation from all directions on all levels.
While a new courtyard space acts as a green social space for the community.
CAR PARKING
EVENT PLAZA
EXISTING BUILDING
LG
L1
L2
L3

August
MAGURO

西北立面 north-west elevation

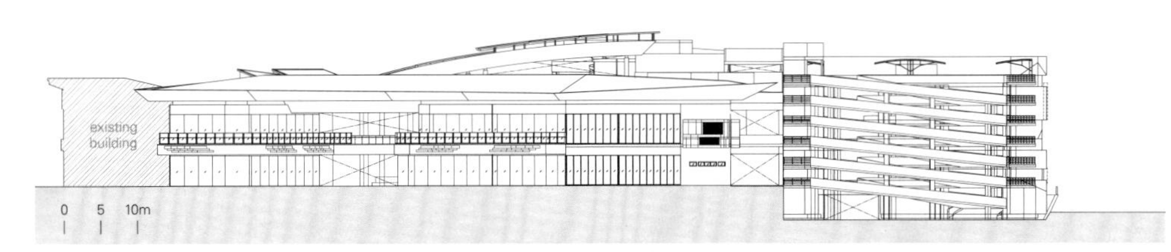

东北立面 north-east elevation

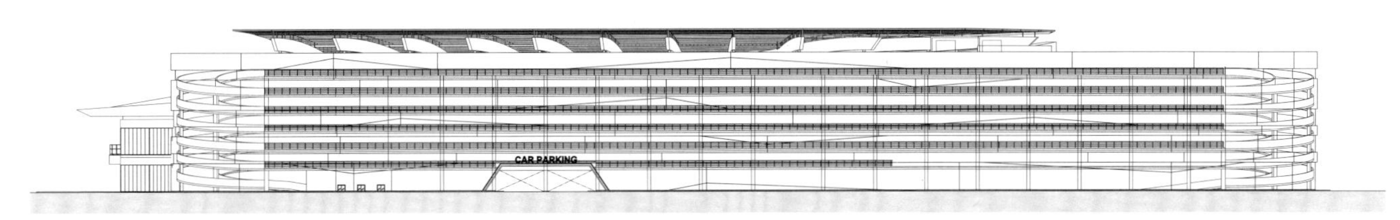

东南立面 south-east elevation

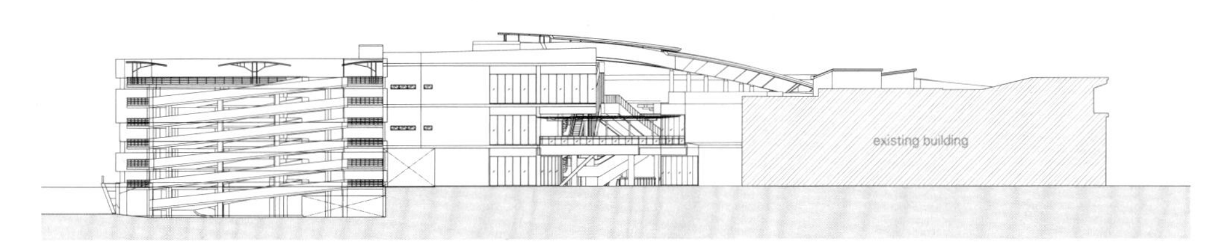

西南立面 south-west elevation

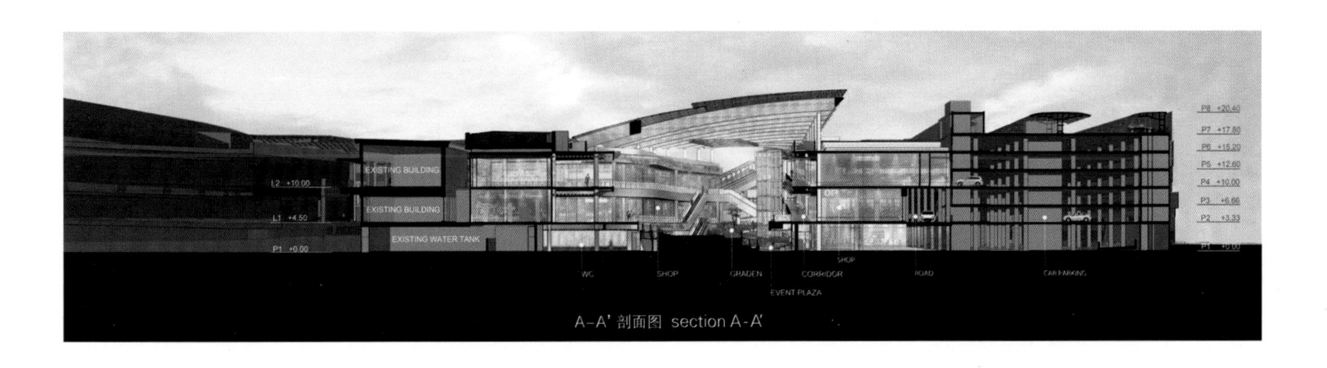

A-A' 剖面图 section A-A'

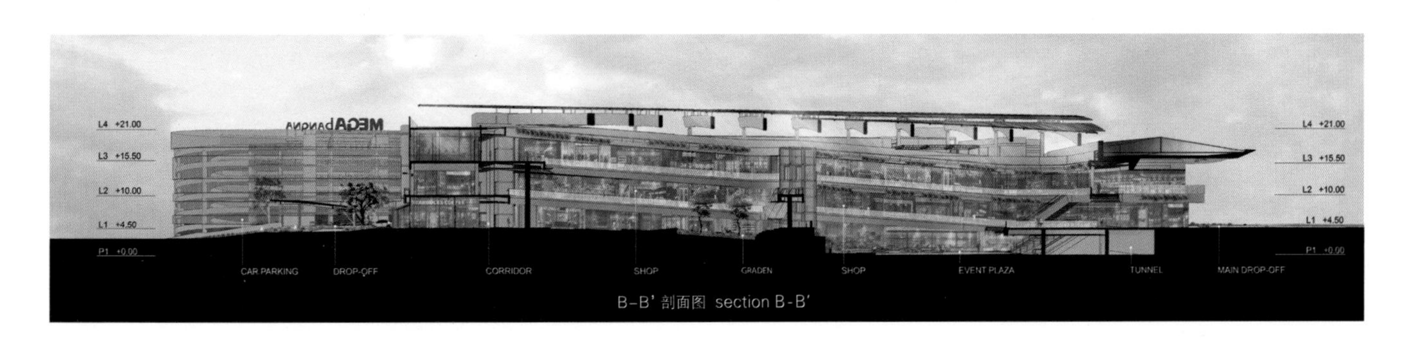

B-B' 剖面图 section B-B'

OISHI EATERIUM

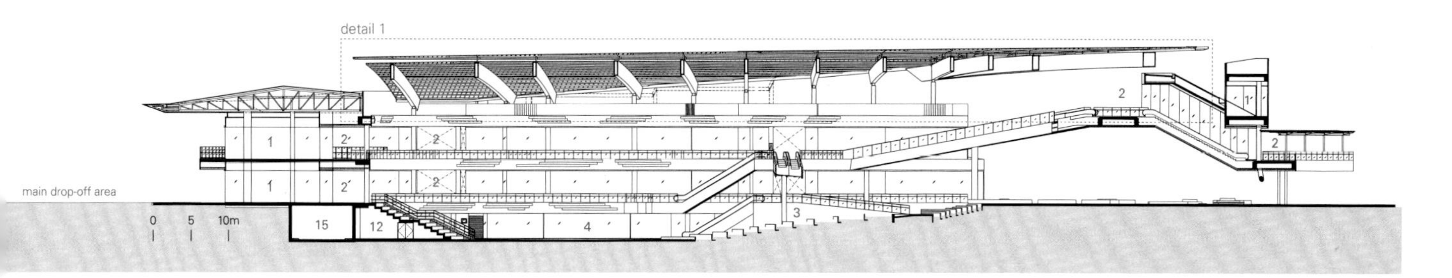

C–C' 剖面图 section C-C'

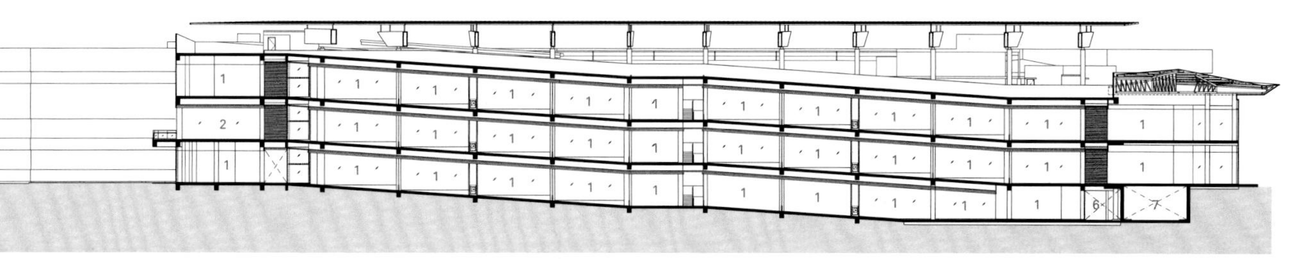

D–D' 剖面图 section D-D'

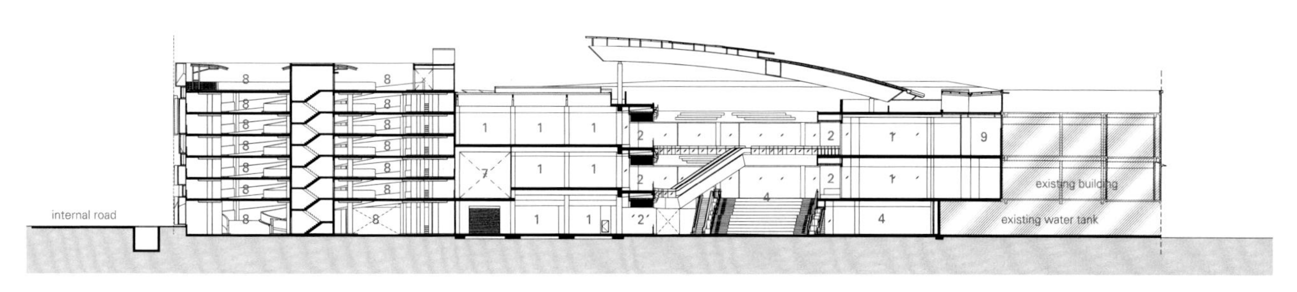

E–E' 剖面图 section E-E'

1. 商店 2. 走廊 3. 花园 4. 活动广场 5. 储藏室 6. 装载区 7. 道路 8. 停车场 9. 服务走廊
1. shop 2. corridor 3. garden 4. event plaza 5. storage room 6. loading area 7. road 8. car parking 9. service corridor

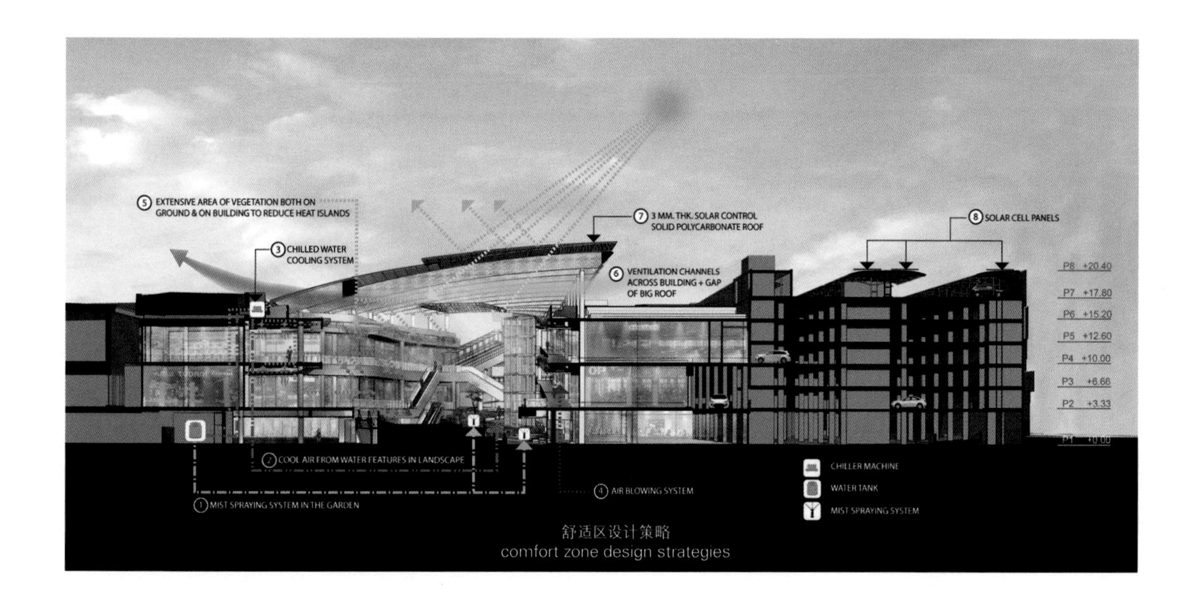

舒适区设计策略
comfort zone design strategies

August
August
WE'RE OPEN
MAGURO
MAGURO

OISHI EATERY
Bonchon

项目名称：Mega Foodwalk / 地点：Megabangna Shopping Complex, Samutprakan, Thailand / 事务所：FOS [Foundry of Space] / 首席建筑师：Rungkit Charoenwat / 指导：Widsara Suthadarat / 项目团队：Rinchai Chaivaraporn, Peera Teerittaveesin, Prapaphan Phongklee, Panintorn Chokprasertthaworn, Pakhon Promta, Singha Ounsakul, Rattha Jiamjarasrangsi, Jirapas Leangsakul, Suvijak Yatinuntsakul / 结构工程师：Aurecon / 机电管道工程师：Aurecon / 景观设计师：Landscape Collaboration / 室内设计师：PIA / 平面设计师：PIA / 照明设计师：WGC International / 成本顾问：RHLB (Siam) Ltd. / 施工经理：Arcadis (Thailand) / 总承包商：Ritta Co., Ltd. / 机电管道承包商：Jardine Engineering Corporation (Thailand) / 客户：SF Development Co., Ltd. / 用地面积：58,000m² / 竣工时间：2018
摄影师：©Rungkit Charoenwat (courtesy of the architect)

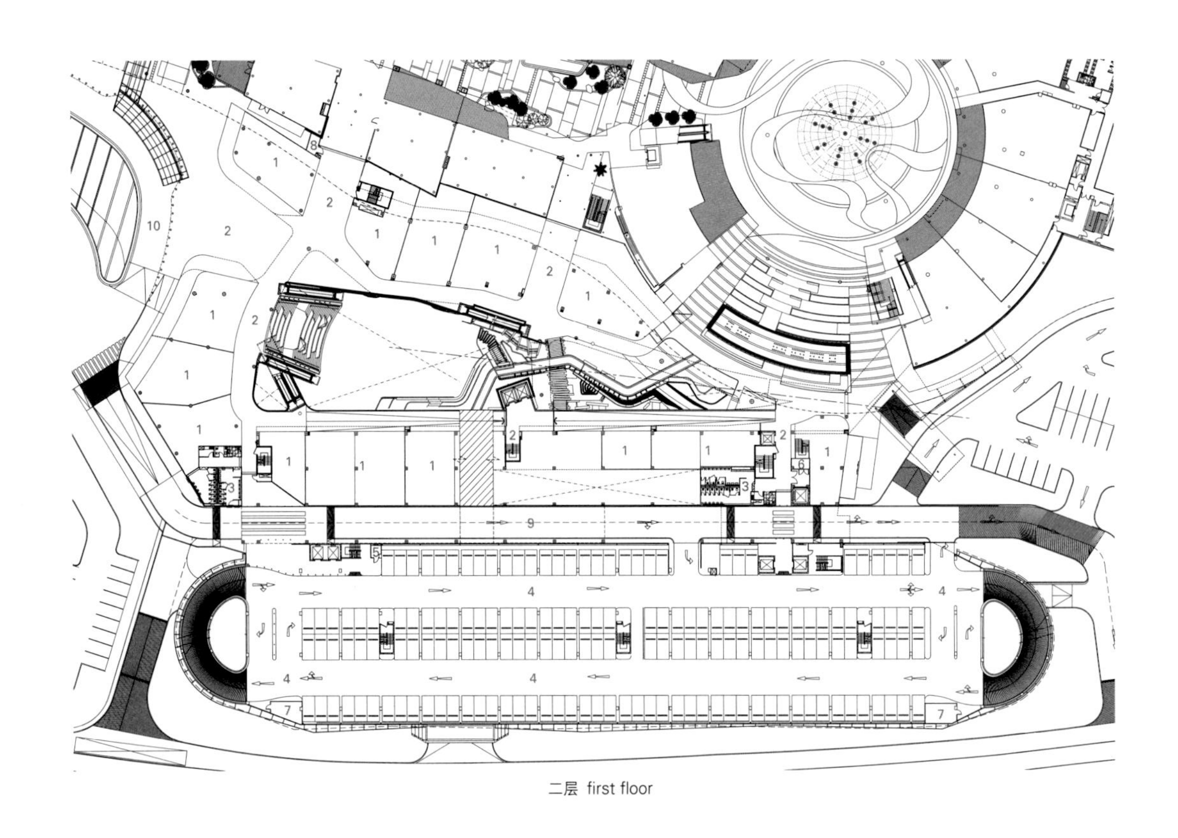

1. 商店
2. 走廊
3. 卫生间
4. 停车场
5. 电气室
6. MDB室
7. 机电管道室
8. 储藏室
9. 道路
10. 乘降点

1. shop
2. corridor
3. wc
4. car parking
5. elec. room
6. MDB room
7. MEP room
8. storage
9. road
10. drop-off

二层 first floor

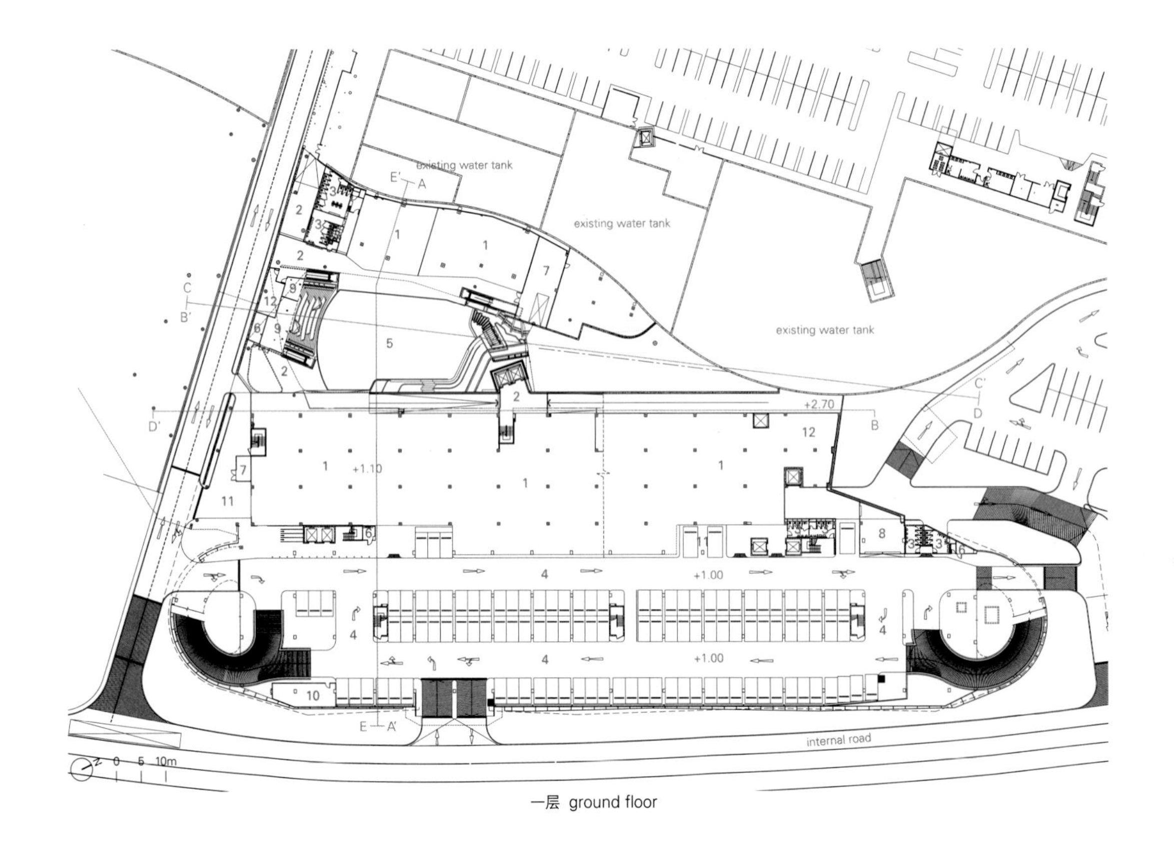

1. 商店
2. 走廊
3. 卫生间
4. 停车场
5. 活动广场
6. 电气室
7. 机电管道室
8. 垃圾房
9. 储藏室
10. 安全室
11. 装载室
12. 更衣室

1. shop
2. corridor
3. wc
4. car parking
5. event plaza
6. elec. room
7. MEP room
8. garbage room
9. storage
10. security room
11. loading area
12. dressing room

一层 ground floor

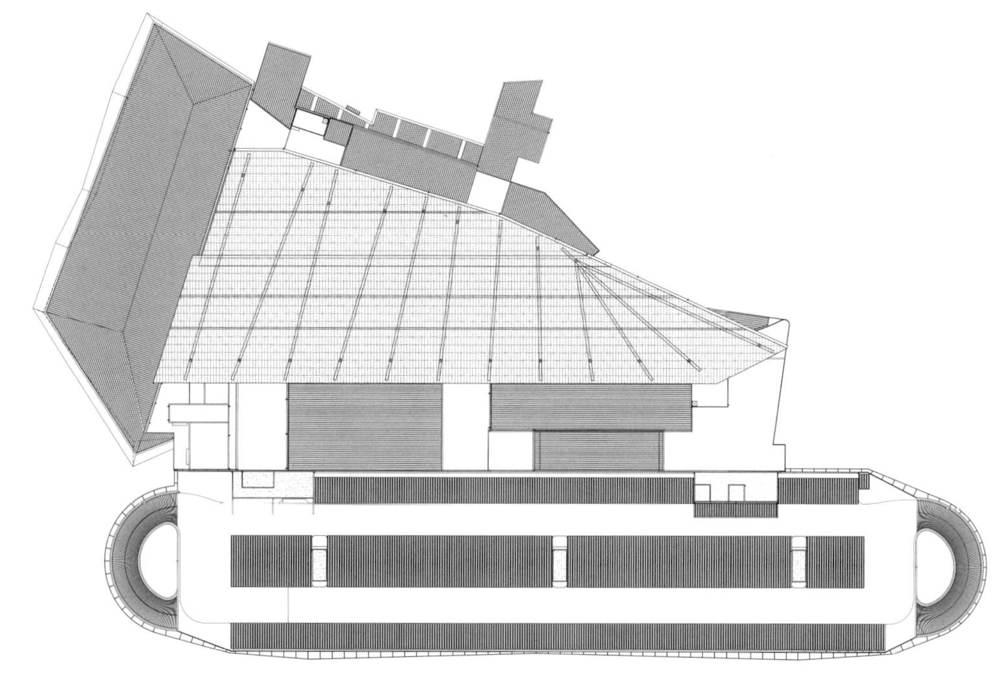

屋顶 roof

1. 商店
2. 走廊
3. 卫生间
4. 停车场
5. 电气室
6. MDB室
7. 机电管道室

1. shop
2. corridor
3. wc
4. car parking
5. elec. room
6. MDB room
7. MEP room

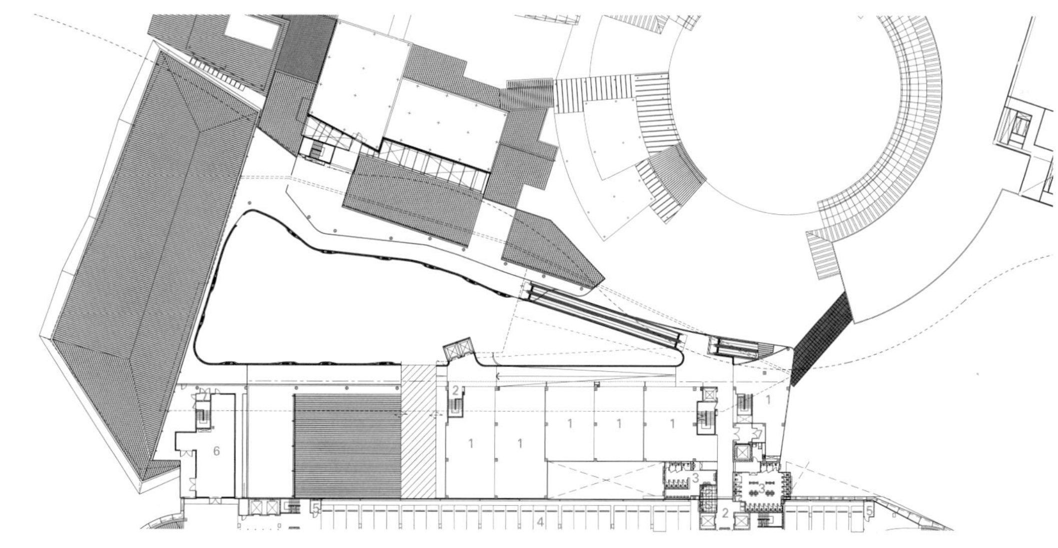

四层 third floor

1. 商店
2. 走廊
3. 卫生间
4. 停车场
5. 电气室
6. 服务走廊

1. shop
2. corridor
3. wc
4. car parking
5. elec. room
6. service corridor

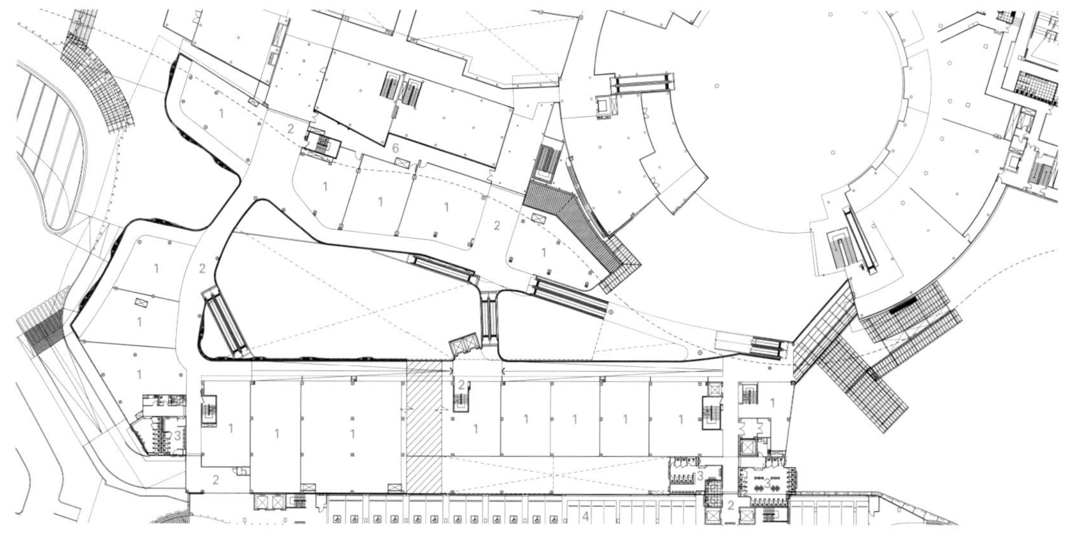

三层 second floor

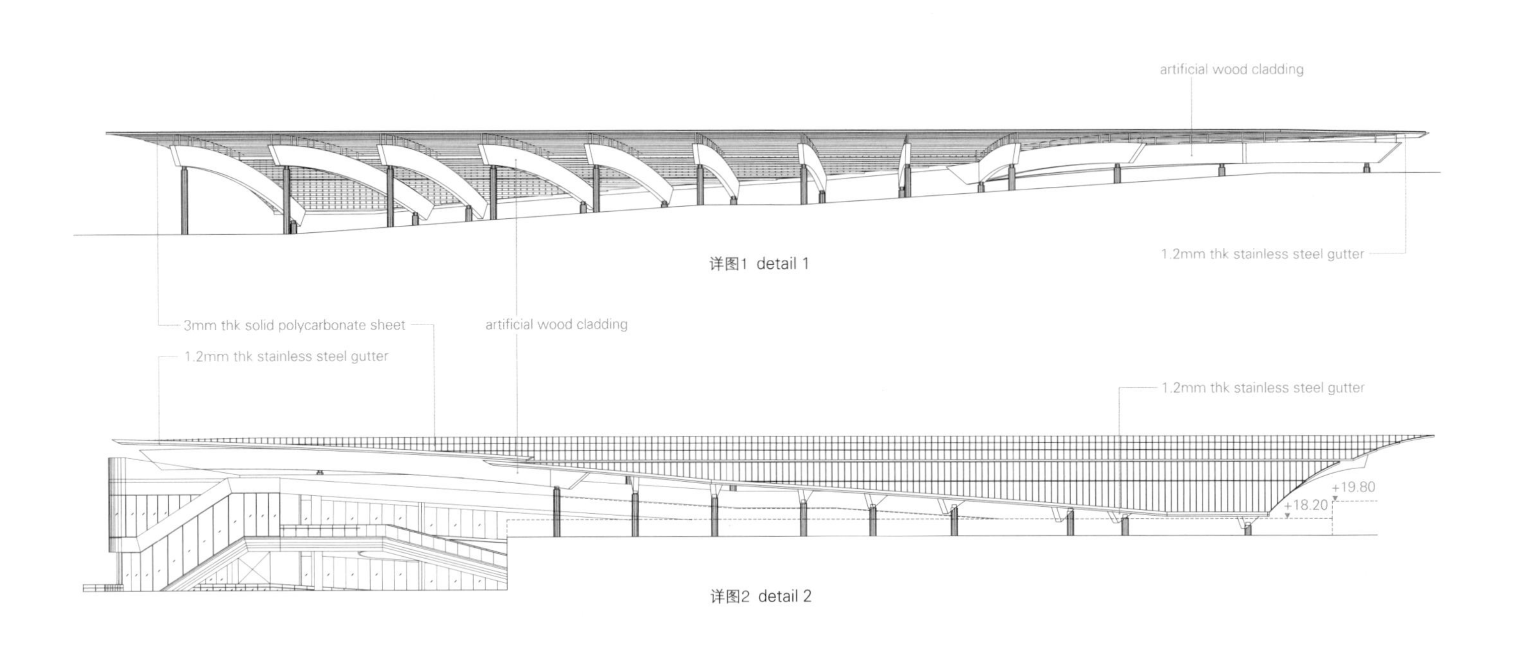

详图1 detail 1

详图2 detail 2

Tops
market

Ribs
Happy Fish

URAZIONI TORINO

苹果自由广场
Apple Piazza Liberty

Foster+Partners

苹果自由广场由两个基本建筑元素构成：阶梯式广场及喷泉。该广场坐落于米兰最受欢迎的步行街之一的Corso Vittorio Emanuele附近。奇特新颖的喷泉吸引了很多游客来到该广场。

广场上标志性的水景是一种互动的、多感官的体验，是对城市生活乐趣的颂扬，体现了城市生活的动感活力。游客可通过一个玻璃入口进入喷泉，感受垂直喷射的水喷射在7.9m高的玻璃墙上所带来的双重感官体验，穿梭于喷泉之中，身临其境一般感受儿时游戏的乐趣。白天，随着阳光洒进喷泉，每每置身其中的感觉也千变万化；夜晚，喷泉喷洒在玻璃墙上，玻璃墙又将波光无限反射到天际，玻璃天花板如万花筒般瞬息万变。

福斯特建筑事务的创始人史蒂芬·贝林评论道："在意大利创建新型公共广场简直是无上光荣并深感责任重大的事情。一直以来，意大利的广场和城市空间设计总能带给我们灵感。在这个设计中，对我们每个人来说，喷泉在诉说着孩子般的兴奋。该设计简洁精炼，想要表达的理念是，走进这个大喷泉而不被淋湿，享受生机勃勃的快乐。"喷泉水向下流入露天剧场基座的下面，露天剧场是一个新的社交中心，也是"今日在苹果"的户外延伸。此露天剧场以第二面喷泉水墙为背景，由宽阔的石阶界定，石阶沐浴在阳光下，延伸到街道下方，通向舞台。整个广场是新建的，铺设的是产于伦巴第的当地经典石材，名为银芝麻，周围栽种了21棵皂荚树。

建筑内部空间明亮、整齐划一，与上方广场构成一个整体，像是雕刻自同一块石头。天花板与圆形剧场阶梯式的设计如出一辙。其中，天窗及背光天花板创意般地将人工灯光与自然光结合在一起。温暖的阳光透过屋顶以及楼梯照射进下沉商店，将其内部与米兰的光线和节奏连接起来，给人一种画廊般宽敞又明亮的视觉感受。抛光不锈钢制成的悬臂阶梯延伸至商店，作为一组灯光雕塑装置，给人一种戏剧性与刺激性结合的双重体验。

Apple Piazza Liberty is an ensemble of two fundamental elements, a stepped plaza and a fountain. Located just off the Corso Vittorio Emanuele – one of the most popular pedestrian streets in Milan – visitors are drawn towards the piazza by the sight of the dramatic new fountain.
A celebration of the joys of city life and embodying its dynamic nature, the signature water feature is an interactive, multisensory experience. Visitors enter the fountain through a glass-covered entrance enveloped by the sights and sounds of vertical jets of water that splash against the

REALE MUTUA DI
OPENING SOON
SQUARE FOOD
café
charleston

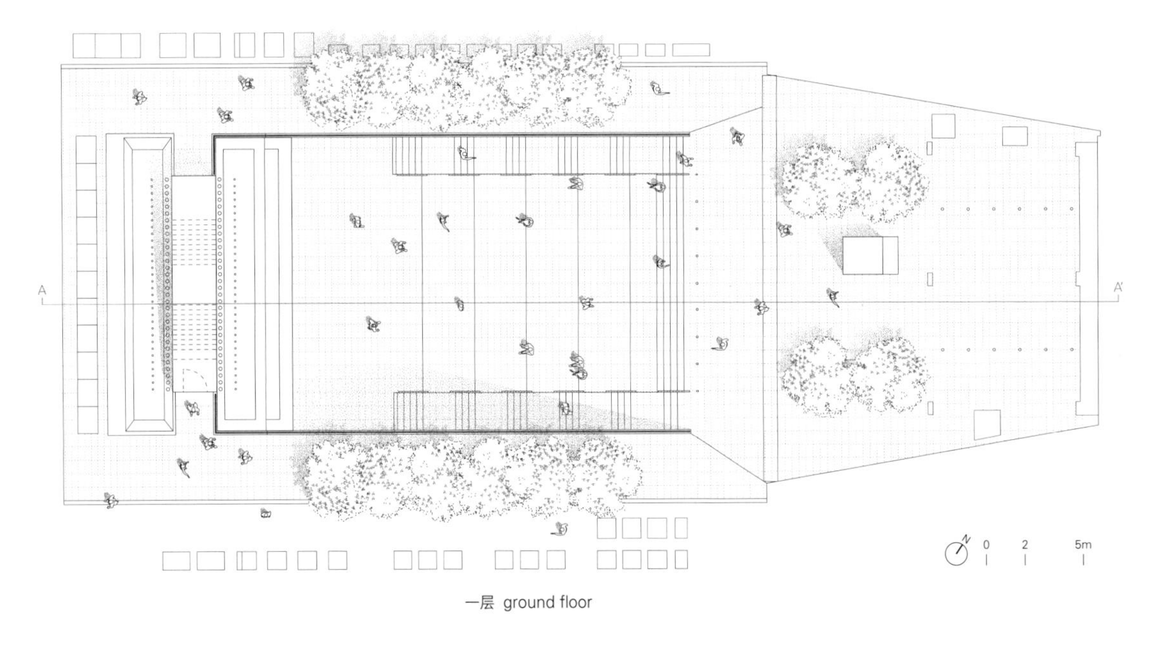

一层 ground floor

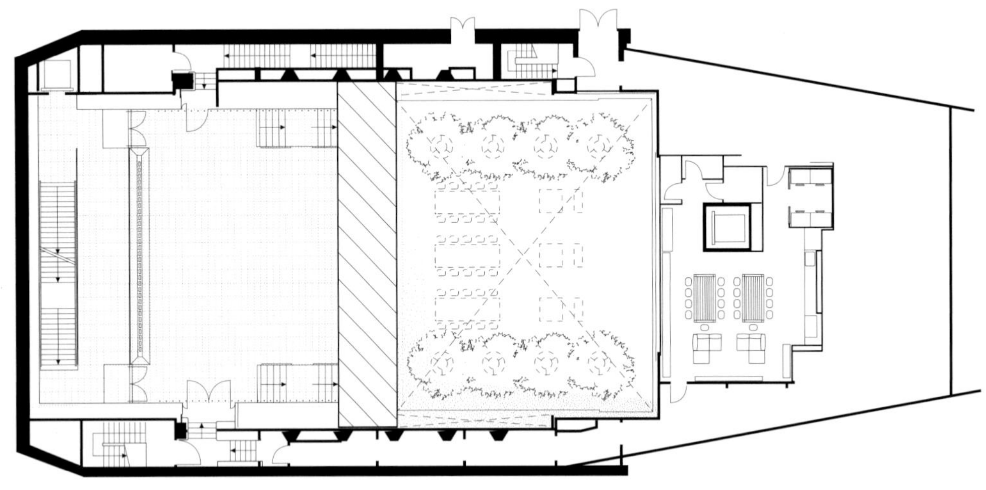

地下一层 first floor below ground

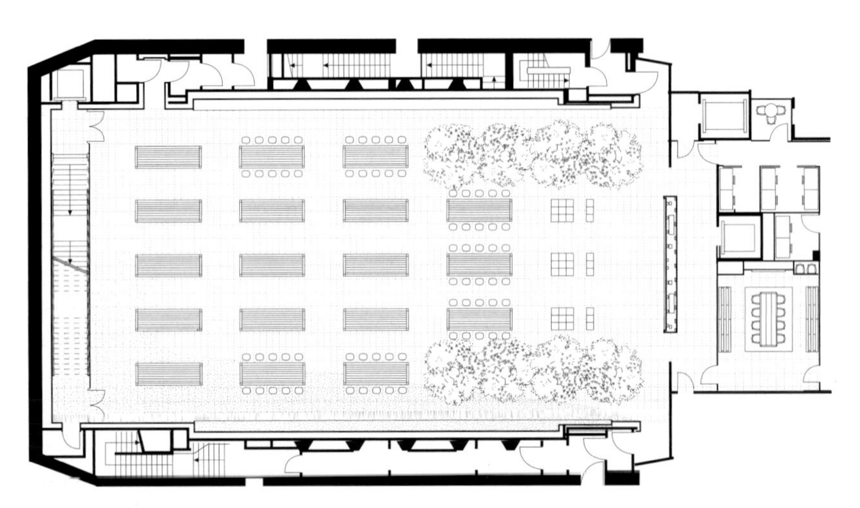

地下二层 second floor below ground

项目名称：Apple Piazza Liberty / 地点：Milan, Italy / 事务所：Foster + Partners / 客户：Apple Inc. / 面积：sales area - 684m²; boardroom - 66m²; total sales area - 940m²; total store - 1,990m²; plaza + amphitheater - 1,750m² / 室内销售区尺寸：height - 4m to 7m, length - 38m, width - 18m, total surface - 684.00m² / 露天剧场尺寸：length - 24.5m, width - 16.5m, stage - 3m below ground, total surface - 405m² / 玻璃结构尺寸：height - 8m, length - 12m, width - 2.5m

东立面 east elevation

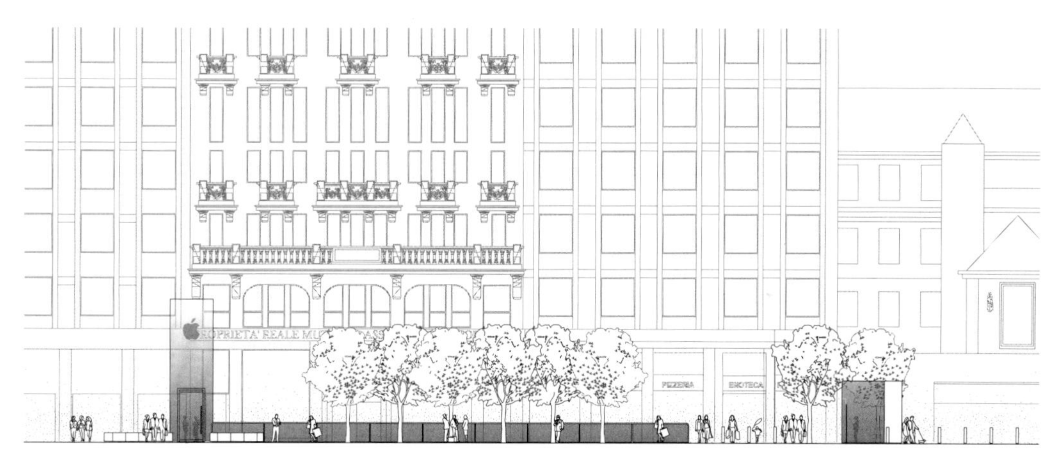

南立面 south elevation

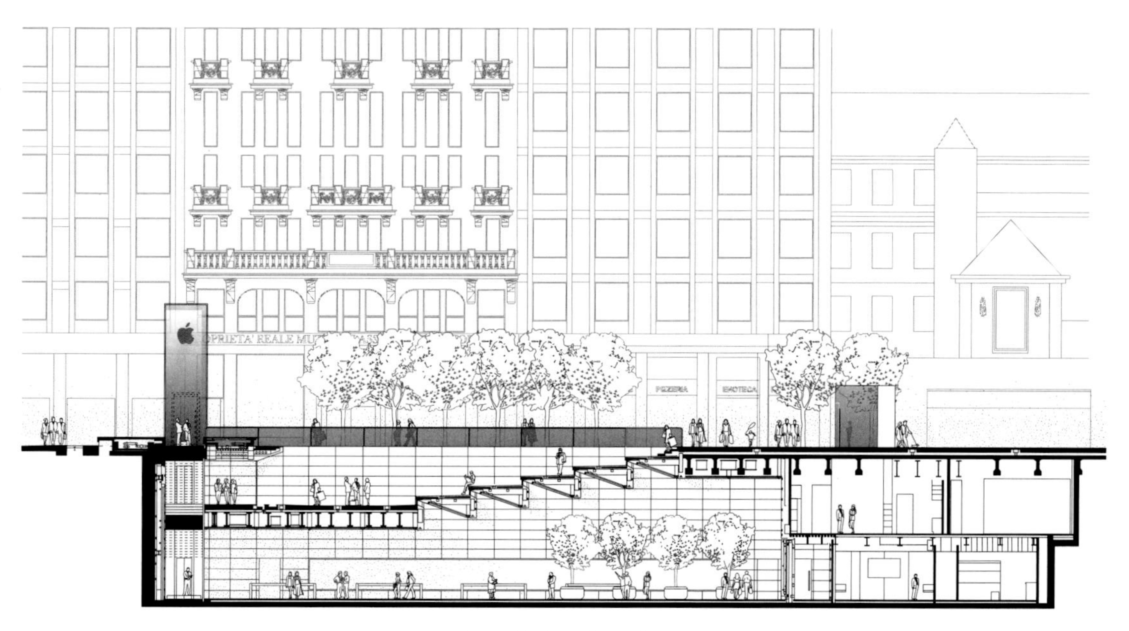

A–A' 剖面图 section A-A'

水景：64 waterjets at upper level, waterjets in fountain reach 8m in height, 4,000 rain droplet nozzles at lower level / 阶梯设计情况：53 step blades, the steps are cantilevered 2.2m from the wall / 石材使用情况：7,776 custom size stones for Interiors, plaza and amphitheatre, 80,227 cobble stones, installed by hand / 委托时间：2014 / 施工开始时间：2016 / 竣工时间：2018 / 摄影师：©Nigel Young (courtesy of the architect)

MARELLA
IRONY su swatch.com

PANINO GIUSTO

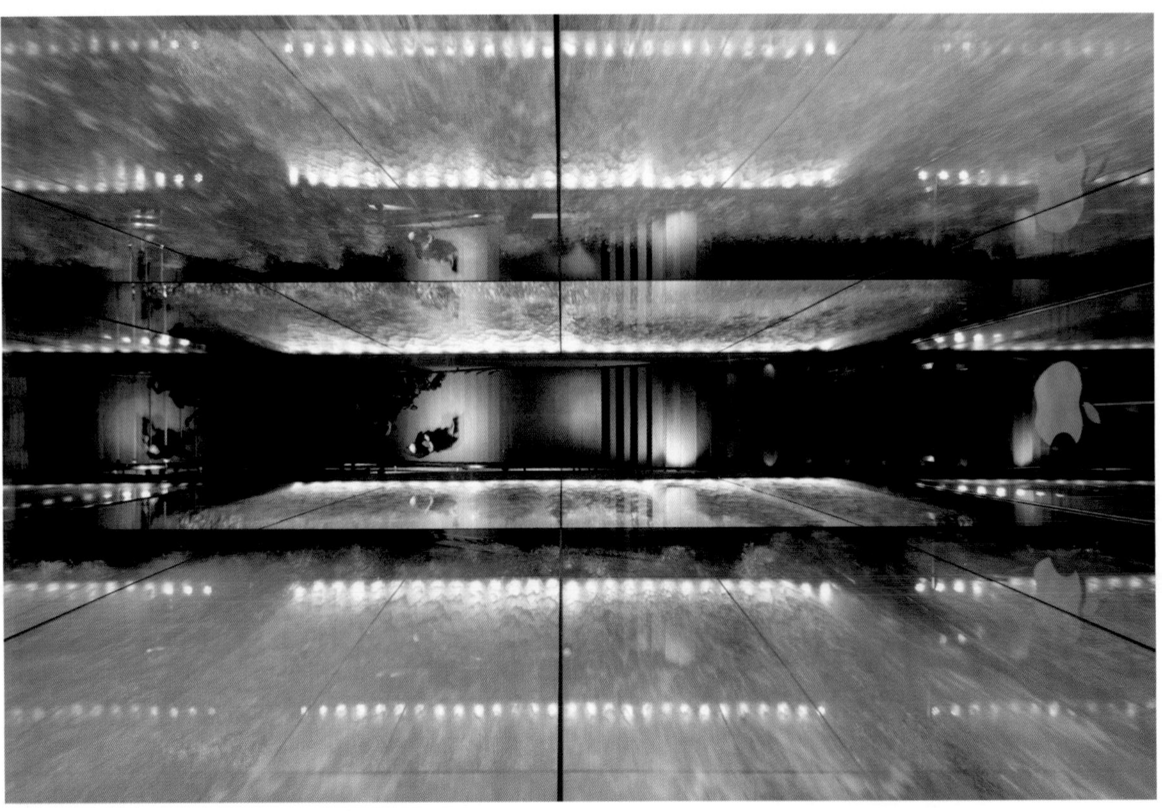

7.9m high glass walls. An immersive recreation of the childhood game of running through fountains, the experience changes throughout the day as sunlight filters through the water, while at night the glass ceiling creates a kaleidoscopic effect, with the water falling down the walls, and its reflections travelling infinitely up the sky.
Stefan Behling, Head of Studio, Foster + Partners, commented: "There can be no greater honor and responsibility than creating a new public plaza in Italy, whose piazzas and urban spaces have always inspired us. The fountain is an expression of child-like excitement that speaks to each one of us. In its simplicity, it echoes the idea of walking into a big fountain without getting wet, and the joy of being alive."
The fountain flows down into the base of the amphitheater, a new social hub and an outdoor extension of "Today at Apple". The amphitheater is defined by broad and sun-soaked stone steps descending below street level and opening up to a stage, backed by a second fountain's wall of water. The entire plaza is newly created and paved with Beola Grigia – a typical local stone from Lombardy, and surrounded by 21 new Gleditsia Sunburst trees.
The interior is a bright, monolithic space, metaphorically carved out of the same stone as the plaza above. The ceiling follows the stepped profile of the amphitheater, with skylights and backlit ceiling panels that innovatively combine artificial and natural light. Through the roof and stairs, warm shafts of sun penetrate deep into the sunken store, connecting the interior with the light and rhythm of Milan and giving it a feel of a spacious daylight-filled art gallery. The stairs leading into the store consist of polished stainless-steel clad cantilevering treads that also become a sculptural light installation, creating a theatrical and exciting experience.

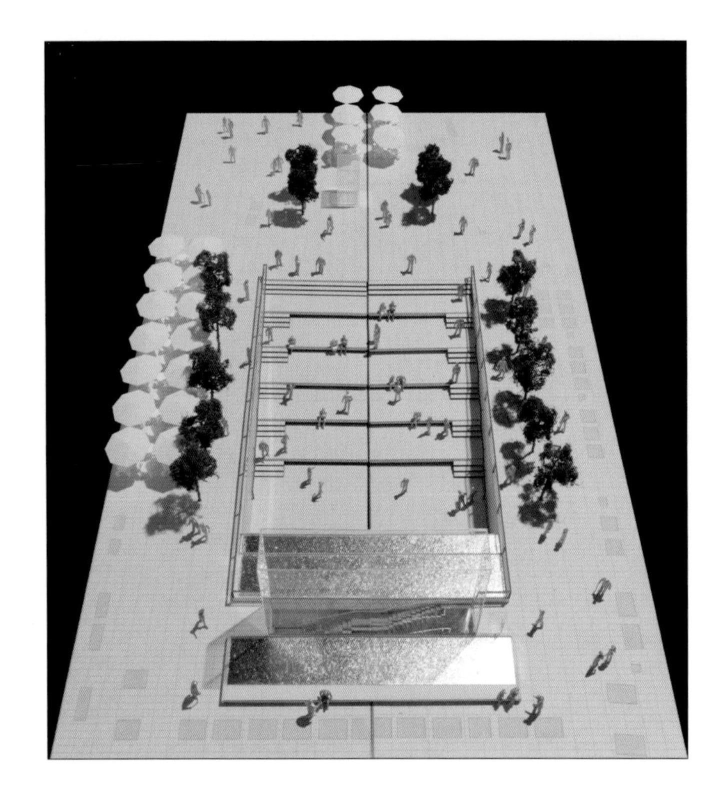

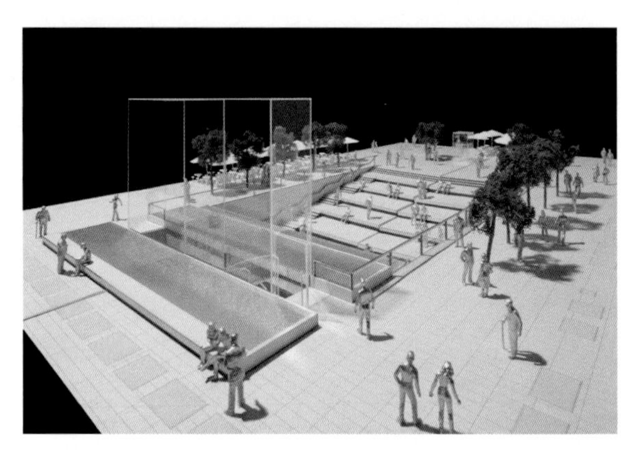

MAX&Co.

阿莫斯·瑞克斯博物馆
Amos Rex Museum

JKMM Architects

HENRY'S PUB
HENRY'S PUB

阿莫斯·瑞克斯博物馆前身是阿莫斯·安德逊艺术博物馆。该博物馆自1965年以来，一直被视为赫尔辛基首屈一指的私人博物馆，展出的是20世纪到21世纪的芬兰艺术品。同时，正如阿莫斯·瑞克斯博物馆馆长卡伊·卡蒂奥所解释的那样，艺术正在从过去"挂在墙上供人瞻仰沉思的东西"，转变为现在"人们一起创造和体验的东西"。随着时间的流逝，阿莫斯·瑞克斯博物馆需要增设一个新的场地，通过一种新形式和新媒体来展示当今日益增多的互动和对话的艺术。

附近的拉西帕拉特西建筑"玻璃宫殿"是芬兰保存最完好的20世纪30年代功能主义建筑之一，现在被确定为博物馆的新家。此建筑最初由三名年轻的芬兰建筑师设计，其定位是要永久矗立于这座城市中。该建筑包括餐厅、商店、电影院和一个露天广场，在20世纪80年代损毁。在赫尔辛基居民保留该建筑免于毁掉的运动之后，JKMM建筑师事务所进行了进一步的保护工作，并设计了阿莫斯·瑞克斯博物馆。JKMM建筑师事务所的设计特别注意保护原有的设计元素，包括建筑立面、门窗、外部霓虹灯照明以及室内灯具在内的一系列设计，重现了与原始设计一致的材料和颜色色调与色彩。

阿莫斯·瑞克斯博物馆的入口设在拉西帕拉特西建筑中，但所有新建的画廊空间都位于地下。在拉西帕拉特西广场下方设计一座占地面积为6230m²的建筑，这一大胆激进的设计来自于JKMM建筑师事务所，同时也是基于该城市的考虑，即拉西帕拉特西广场将如何继续成为一个重要的市民空间，同时，也能让公众看到阿莫斯·瑞克斯博物馆唯一可见的新建筑立面。解决方案是采用高大的雕塑般屋顶天窗的形式，这样做也解决了将阳光引入地下展览空间的问题。屋顶天窗的设计创造了一个新的起伏地形，其和缓绵延起伏的形状犹如一座城市公园，保持了广场的完整性。穹顶的形状构成这一新建公共广场的地形，此公共广场的下面就是新建的画廊空间。整个穹顶缓缓起伏的表面由混凝土地砖铺设，共铺设了13000m³混凝土砖。

从翻新后的拉西帕拉特西的门厅进入，参观者走下楼梯，经过一扇可看到公共广场优美景色的窗户，然后进入2200m²的地下画廊。地下画廊为无柱空间，因此可容纳大规模艺术品展览。在这个灵活度非常大的空间内可进行展览、放置设施以及表演，这里具有高度的技术可控性。在画廊内部，参观者仰望由宽大钢框混凝土制成的天窗，可以看到通过对天窗的角度精心设计才能看到的精挑细选的室外景观，看到外部街道，参观者因此会感觉与城市连为一体。

阿莫斯·瑞克斯博物馆的展览项目，从最新、实验性的当代艺术品，到20世纪现代主义及古代文化，致力于以一种清新充满活力的姿态展现高级宏伟的艺术，致力于无论在过去、现在，还是未来，都能给人带来独一无二的体验；无论在地上、地下，还是在屏幕上，都能创造出令人惊奇的邂逅。拉西帕拉特西广场原有的餐厅及商店继续营业，为公共广场的生活和活动提供便利。这是赫尔辛基市中心社交生活的焦点，也是购物与娱乐场所最重要的公共空间之一。此广场也为阿莫斯·瑞克斯博物馆提供了举办户外活动的机会，为室内展览提供支持。加上拉西帕拉特西展馆，现在阿莫斯·瑞克斯博物馆综合体达到了13 000m²。从建筑以及文化的角度来看，赫尔辛基不断发展的城市形象对构思阿莫斯·瑞克斯博物馆这一项目起到至关重要的作用。这是一个真正令人振奋的视觉艺术新中心。

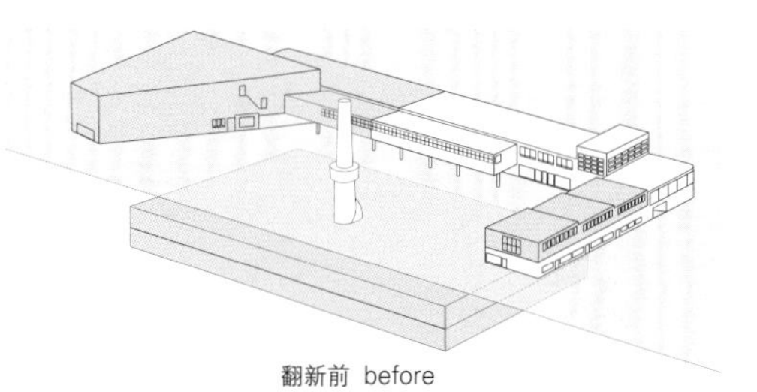
翻新前 before

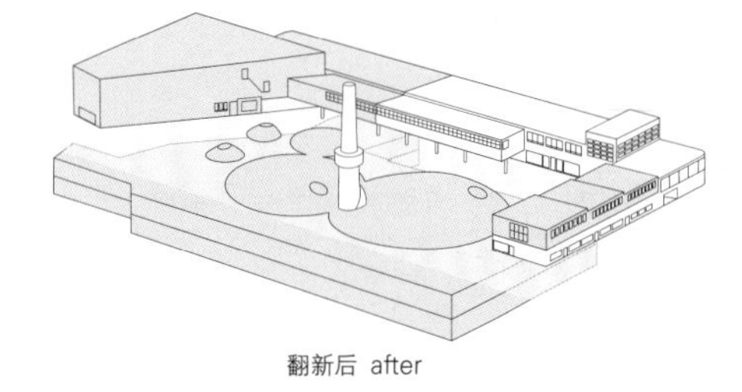
翻新后 after

拉西帕拉特西广场，1937年
Lasipalatsi, 1937

拉西帕拉特西广场，1952年
Lasipalatsi, 1952

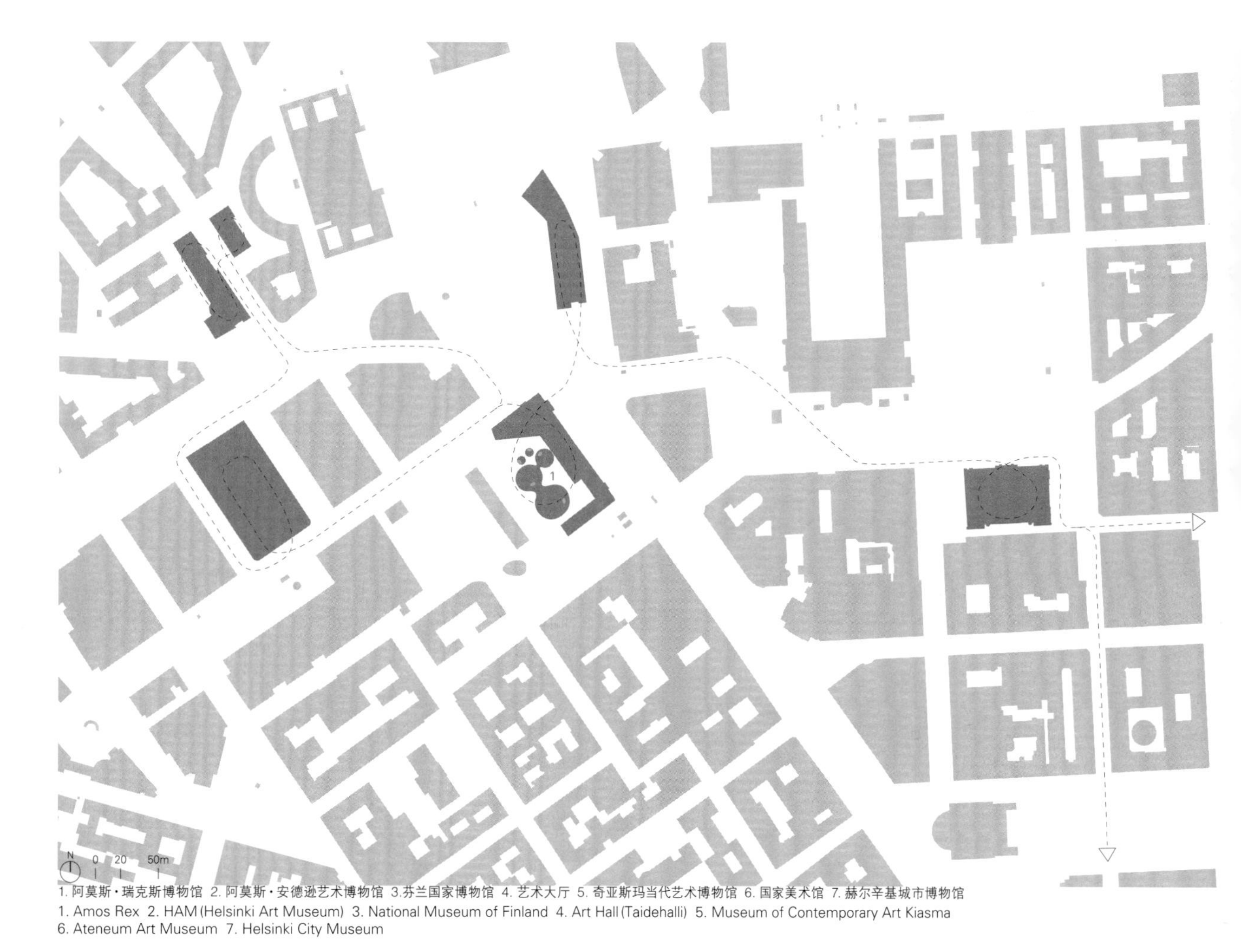

1. 阿莫斯·瑞克斯博物馆 2. 阿莫斯·安德逊艺术博物馆 3.芬兰国家博物馆 4. 艺术大厅 5. 奇亚斯玛当代艺术博物馆 6. 国家美术馆 7. 赫尔辛基城市博物馆
1. Amos Rex 2. HAM (Helsinki Art Museum) 3. National Museum of Finland 4. Art Hall (Taidehalli) 5. Museum of Contemporary Art Kiasma 6. Ateneum Art Museum 7. Helsinki City Museum

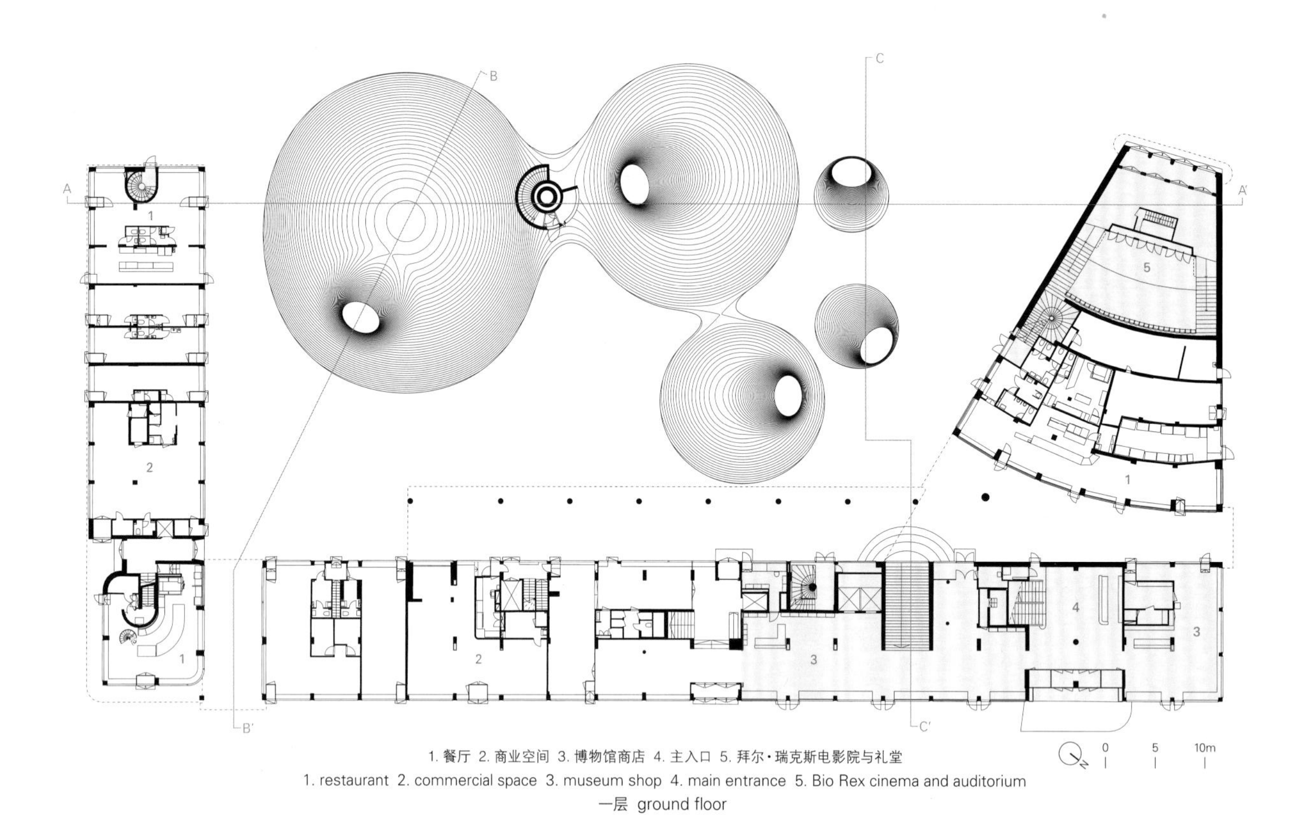

1. 餐厅 2. 商业空间 3. 博物馆商店 4. 主入口 5. 拜尔·瑞克斯电影院与礼堂
1. restaurant 2. commercial space 3. museum shop 4. main entrance 5. Bio Rex cinema and auditorium
一层 ground floor

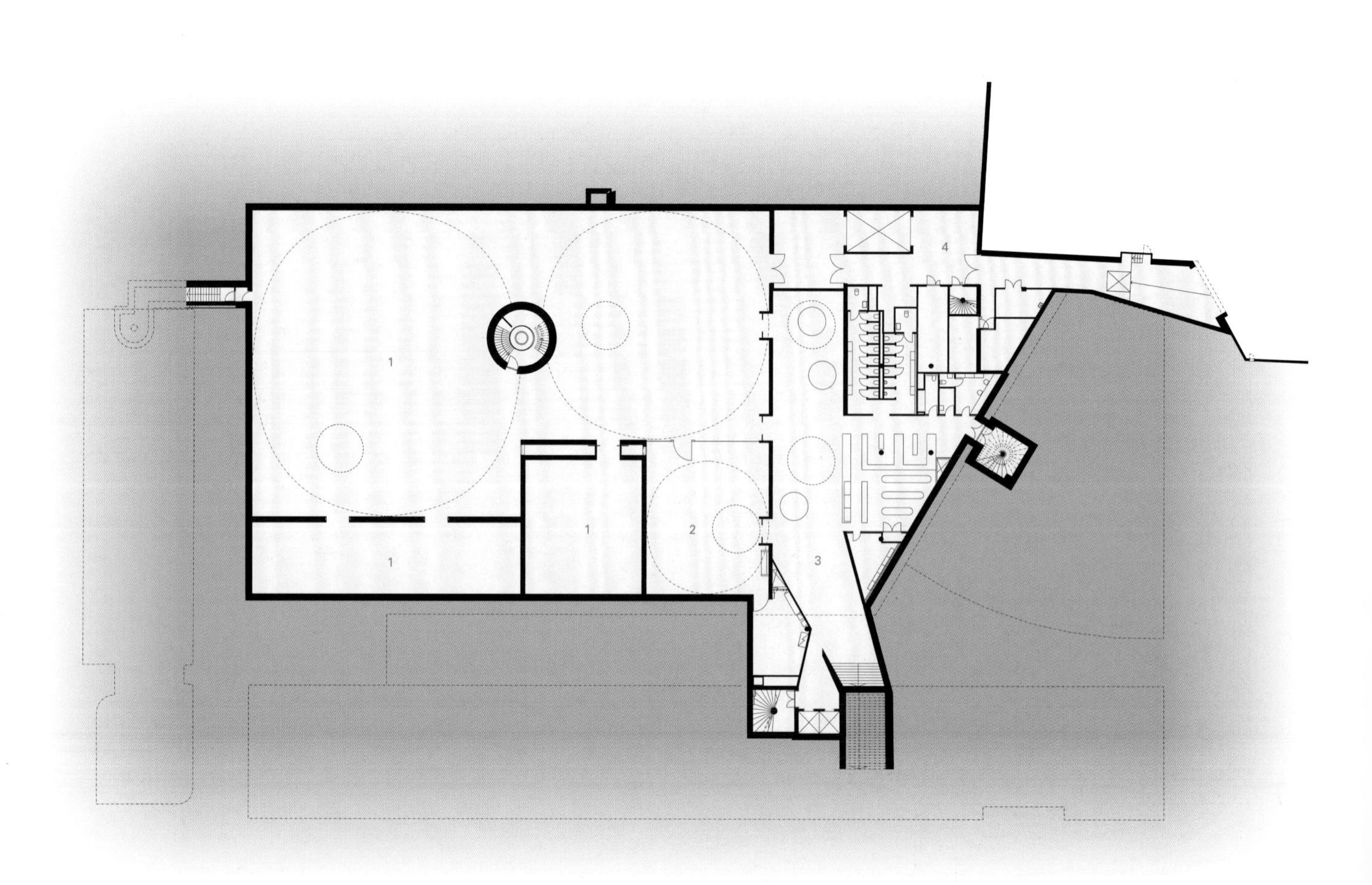

1. exhibition space 2. workshop 3. lobby 4. service
地下一层 first floor below ground

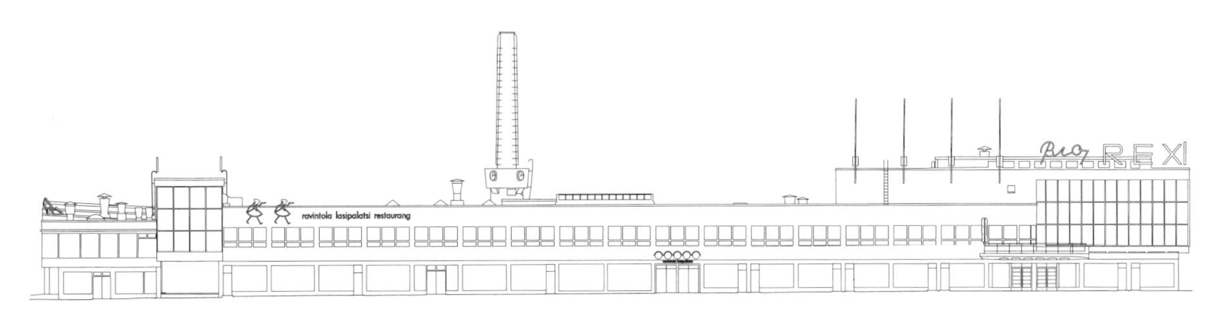

东北立面 north-east elevation

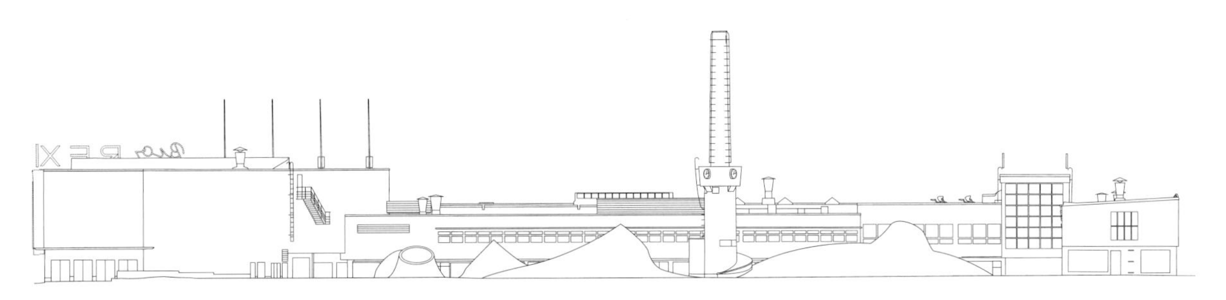

西南立面 south-west elevation

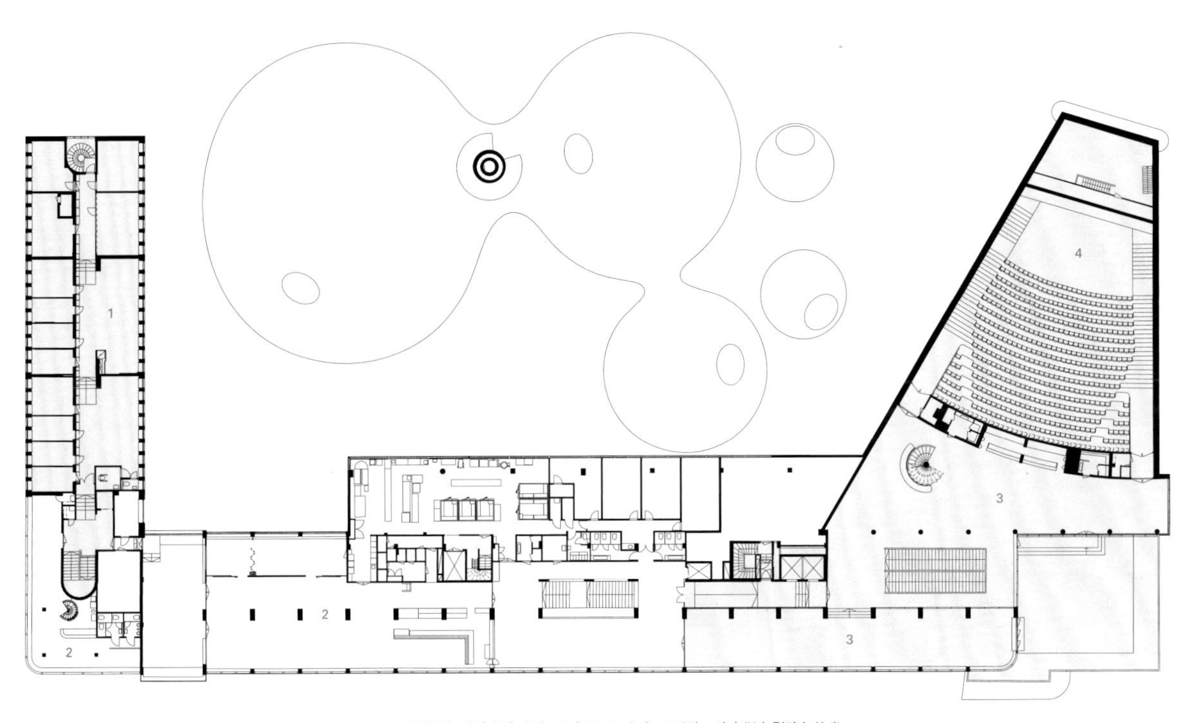

1. 阿莫斯·瑞克斯办公室 2. 餐厅 3. 大堂 4. 拜尔·瑞克斯电影院与礼堂
1. Amos Rex offices 2. restaurant 3. lobby 4. Bio Rex cinema and auditorium
二层 first floor

Amos Rex has its origins in the Amos Anderson Art Museum, which has been regarded as Helsinki's leading private museum showcasing 20th and 21st century Finnish art since 1965. Meanwhile, art is changing from "something you hang on the wall and go respectfully to contemplate", as explained by the director of Amos Rex, Kai Kartio, to "something people make and experience together". In the stream of time, it requires a new venue for providing today's increasingly interactive and conversational art in new forms and new media.

The nearby Lasipalatsi building, "glass palace", one of the best-preserved examples of 1930s functionalist Finnish architecture, was identified as a new home for the museum. Originally designed by three young Finnish architects, it was positioned permanent in the city with its restaurants, shops, a cinema, and an open square, yet fell into disrepair in the 1980s. After Helsinki residents' campaign to save the building from demolition, JKMM Architects followed with a further conservation works and designed the Amos Rex Museum. Special care was given to preserve original features including the building's facades, doors and windows, external neon lighting and interior light fixtures to recreate a palette of materials and colors that are true to the original design.

The entrance to Amos Rex is through Lasipalatsi. All the new gallery spaces of Amos Rex, however, are underground. This radical approach of designing a 6,230m² building below the Lasipalatsi Square was based on JKMM and the city's consideration on how Lasipalatsi Square would remain as an important civic space while also allowing the public to

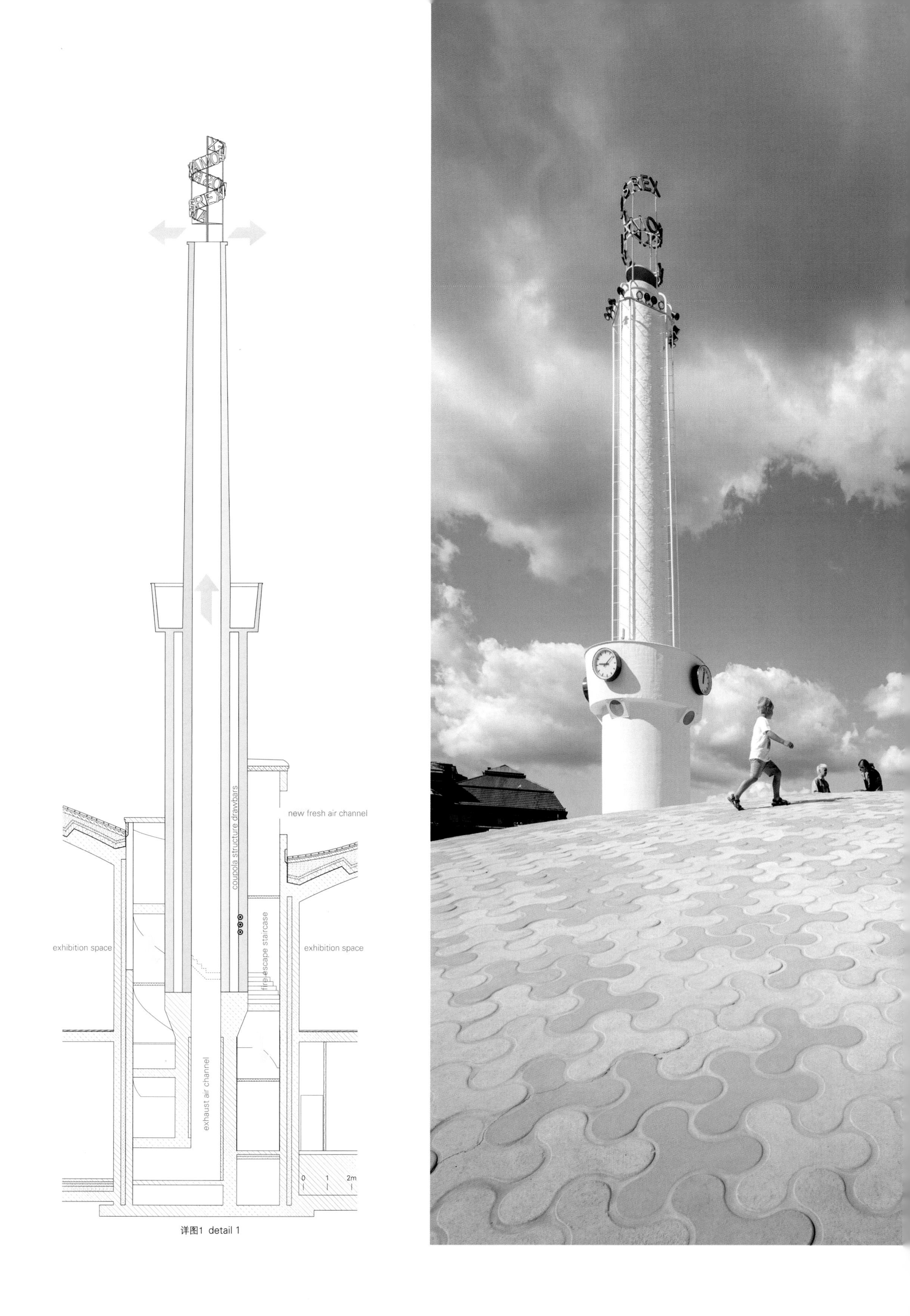
new fresh air channel
coupola structure drawbars
exhibition space
fire escape staircase
exhibition space
exhaust air channel
0 1 2m

详图1 detail 1

enjoy Amos Rex's only visible newly built elevation. The solution came in the form of highly sculptural roof lights that also address the challenge of bringing daylight into the subterranean exhibition spaces. The roof lights create a new undulating topography, the gently rolling forms playing on the idea of an urban park in keeping the integrity with the square. The shape of the domes is expressed in the topography of the newly landscaped public square which sits above the galleries, as a series of gently rolling forms clad in concrete tiles – 13,000m^3 of rock.

From the foyer of the refurbished Lasipalatsi, visitors descend a staircase past a picture window that affords views over the public square, into the basement galleries of 2,200m^2, which are made column-free thus have the opportunity to accommodate large-scale artworks, and to stage exhibitions, installations and performance in a hugely flexible space with a high degree of technical control. Inside the gallery spaces, visitors looking up to the generous steel-framed concrete skylights will feel connected to the city through carefully considered views opening out to the street-level above.

Amos Rex's exhibition programme extends from the newest, often experimental, contemporary art to 20th-century Modernism and ancient cultures. It aims to present captivating and ambitious art refreshingly and exuberantly. The goal is for the past, the present, and the future to produce unique experiences and surprising encounters beneath and above ground, and on the screen. The existing restaurants and shops within the Lasipalatsi will continue to help contribute to the life and activity of the public square, which is a focal point for the social life of Helsinki city center and one of the most important public spaces in the shopping and entertainment district. The square also provides an opportunity for Amos Rex to create a programme of outdoor events to support its gallery shows. Together with the Lasipalatsi Pavilion, the museum complex extends to 13,000m^2. From an architectural and a cultural perspective, Helsinki's evolving urban identity has been paramount in conceiving the Amos Rex project, a truly exciting new center for the visual arts.

the athlete
TRANS
Cubus

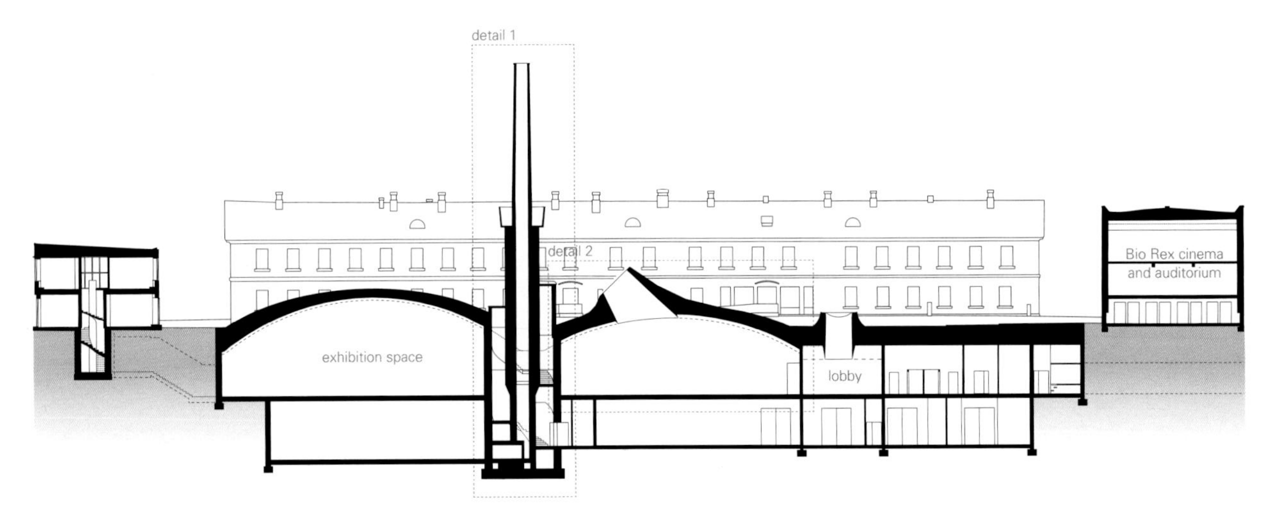

A-A' 剖面图 section A-A'

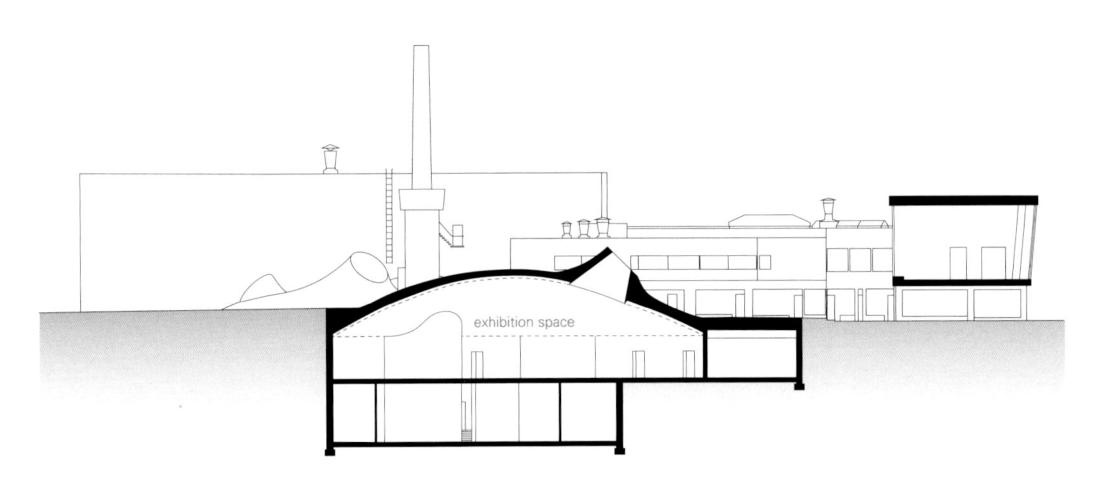

B-B' 剖面图 section B-B'

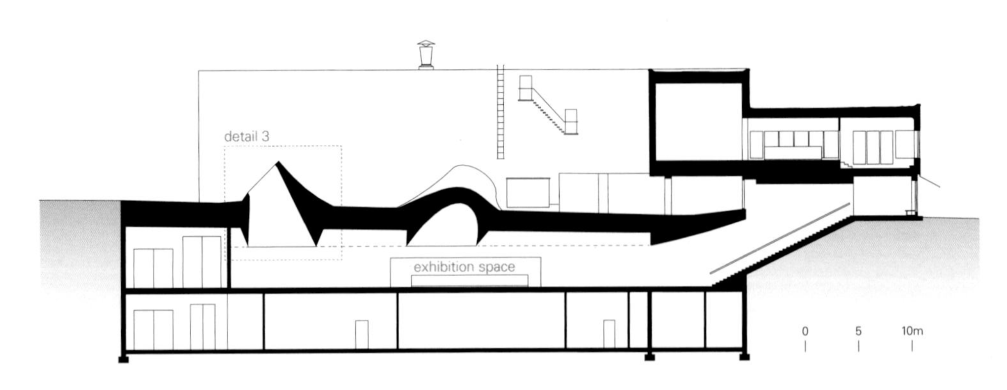

C-C' 剖面图 section C-C'

项目名称：Amos Rex Museum / 地点：Mannerheimintie 22–24, FI–00100 Helsinki, Finland / 事务所：JKMM Architects / 项目团队：Asmo Jaaksi-lead architect, architect SAFA; Freja Ståhlberg-Aalto project architect, architect SAFA; Katja Savolainen-project architect, architect SAFA; Teemu Kurkela, architect SAFA; Samuli Miettinen, architect SAFA; Juha Mäki-Jyllilä, architect SAFA; Edit Bajsz, architect SAFA; Christopher Dolany, architect; Markus Manninen, architect SAFA; Marko Pulli, architect; Katariina Takala-BIM coordinator; Jarno Vesa-interior architect; Jussi Vepsäläinen, architect, Päivi Meuronen interior architect SIO; Noora Liesimaa-interior architect SIO / 项目管理：Haahtela-rakennuttaminen Oy / 结构设计：Sipti Oy Structural design of the domes: Sweco Rakennetekniikka Oy / 暖通空调工程师：Ramboll Talotekniikka Oy / 电气工程师：Ramboll Talotekniikka Oy / 穹顶结构设计：Ins.tsto Heikki Helimäki Oy, Helimäki Akustikot / 防火顾问：L2 Paloturvallisuus Oy / 客户：Föreningen Konstsamfundet, The Amos Anderson Art Museum, The City of Helsinki / 用途：Galleries, office and supporting back office spaces, commercial, technical and logistical spaces / 用地面积：13,000m² / 建筑面积：2,200m² / 材料：Lasipalatsi square-unique concrete tiles; Exhibition halls, ceilings-textile upholstered, perforated metal discs; Exhibition halls, floorings-block flooring, finnish pine; Lobby, ceiling-textile lightning fixture; Lobby, flooring-white concrete / 造价：€50 million / 结构时间：2018 / 摄影师：©Tuomas Uusheimo (courtesy of the architect)-p.63[right], p.66, p.67, p.70~71, p.72~73, p.75; ©Mika Huisman (courtesy of the architect)-p.60~61, p.62~63[bottom], p.68~69, p.76~77

1. landscape paving
   80mm concrete tile, unique "clover" shape
   impermeable cement-based jointing mortar
   50mm cement-based/stone dust bed
   120mm reinforced concrete slab
   >150mm glass foam filling and insulation
   300/400mm polystyrene thermal insulation
   drainage sheet
   three-ply bitumen sheeting
2. reinforced concrete cupola structure
   200mm general construction
3. pre-stressed reinforced concrete beam structure enveloping each cupola, restraining horizontal forces
4. asphalt paving
   40mm + 40mm asphalt coating
   120mm reinforced concrete slab
   400mm polystyrene thermal insulation
   drainage and three-ply bitumen sheeting
   400mm prestressed reinforced concrete slab
5. suspended ceiling
   perforated aluminum discs, 5mm, d-440mm, attached to each other by C-profile wings, 50mm
   fireproof textile hoods
   technical installation space 150mm for sprinkler, fire sensor, power and data systems
   acoustic board, 50mm
6. lighting tracks, grid of 2.7m x 2.7m, power and data channels
7. suspension points, grid of 2.7m x 2.7m, load bearing 3kN (300kg)
8. sprinkler system, heads by grid of 2.7m x 2.7m
9. drainage

详图2 detail 2

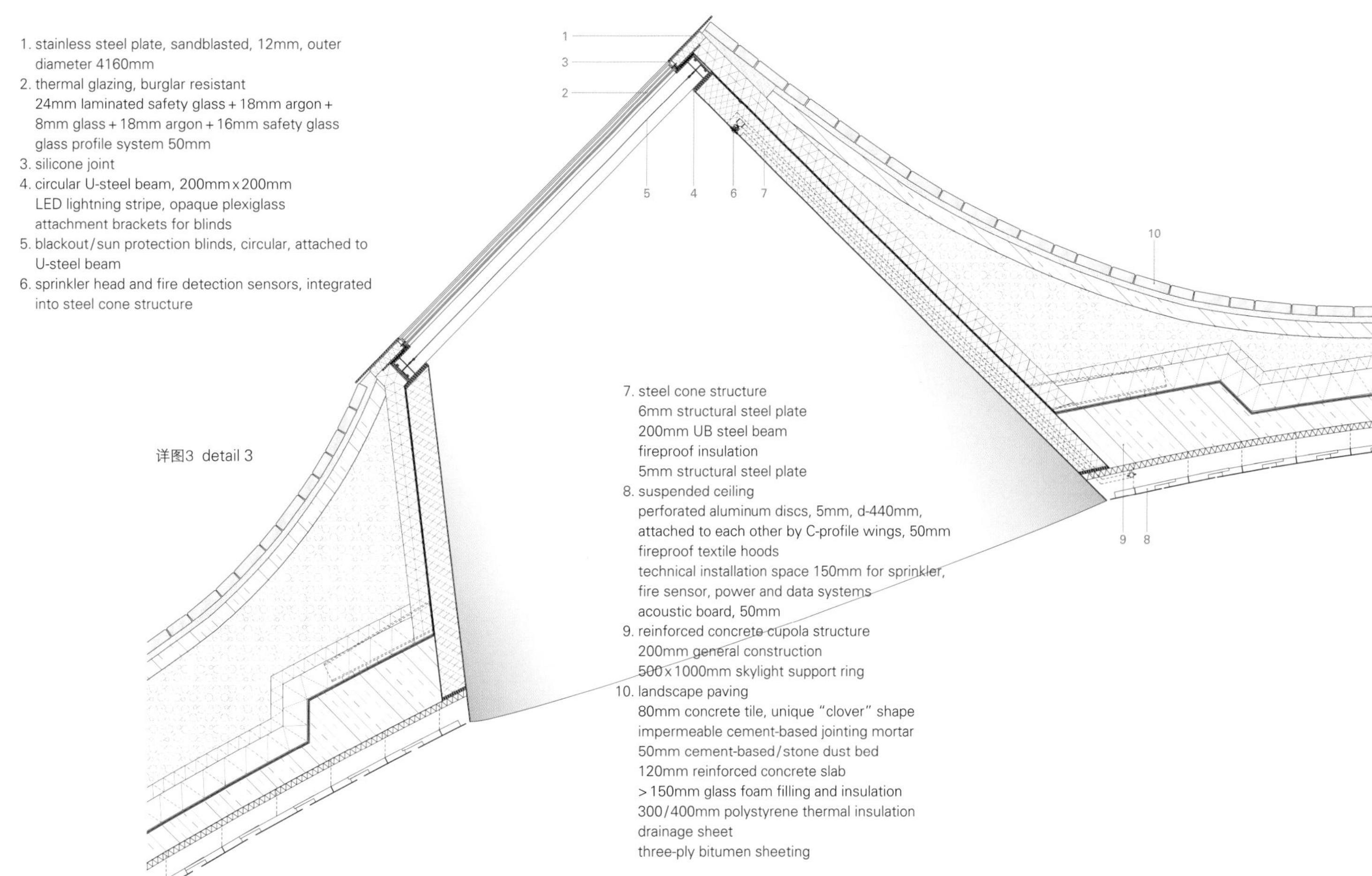

1. stainless steel plate, sandblasted, 12mm, outer diameter 4160mm
2. thermal glazing, burglar resistant
   24mm laminated safety glass + 18mm argon + 8mm glass + 18mm argon + 16mm safety glass
   glass profile system 50mm
3. silicone joint
4. circular U-steel beam, 200mm x 200mm
   LED lightning stripe, opaque plexiglass
   attachment brackets for blinds
5. blackout/sun protection blinds, circular, attached to U-steel beam
6. sprinkler head and fire detection sensors, integrated into steel cone structure
7. steel cone structure
   6mm structural steel plate
   200mm UB steel beam
   fireproof insulation
   5mm structural steel plate
8. suspended ceiling
   perforated aluminum discs, 5mm, d-440mm, attached to each other by C-profile wings, 50mm
   fireproof textile hoods
   technical installation space 150mm for sprinkler, fire sensor, power and data systems
   acoustic board, 50mm
9. reinforced concrete cupola structure
   200mm general construction
   500 x 1000mm skylight support ring
10. landscape paving
   80mm concrete tile, unique "clover" shape
   impermeable cement-based jointing mortar
   50mm cement-based/stone dust bed
   120mm reinforced concrete slab
   > 150mm glass foam filling and insulation
   300/400mm polystyrene thermal insulation
   drainage sheet
   three-ply bitumen sheeting

详图3 detail 3

阿莫斯·瑞克斯博物馆：赫尔辛基的新文化动力源

首席建筑师阿斯莫·亚克思与修复工程建筑师卡地亚·萨博拉宁访谈

Amos Rex: Helsinki's New Cultural Powerhouse

Interview with Asmo Jaaksi, principal architect, and Katja Savolainen, restoration architect

采访人：拉西帕拉特西展馆本打算用于举办临时展览，但意想不到的是，竟一直开放，并开放了80多年，甚至被国际现代建筑文献组织认为是芬兰现代主义杰作之一。

阿斯莫·亚克思：原拉西帕拉特西建筑是三位年轻建筑师Kokko、Revell、Riihimäki的作品。即使在今天，也能在这个建筑作品上看到三位建筑师的青春热情、卓越的技能和发自内心的激情。不过，该设计与时代紧密相关。拉西帕拉特西是芬兰功能主义早期作品，甚至还颇有创见。因为此建筑与建于19世纪并列入文物保护名录的新古典主义风格的营房相邻，所以作品本身也是在这样比较敏感的历史背景下完成的。营房被列入文物保护名录是因为保留了以前的阅兵场，也就是现在的拉西帕拉特西广场，它仍是一个没有任何建筑物的开放式公共空间。现在，阿莫斯·瑞克斯博物馆建于其下也是为了保护这一开放式公共空间。

有趣的是，最初是把拉西帕拉特西设计为一个多功能、多用途的展馆类型的设施的，设计有餐厅、商店和电影院，为1940年在赫尔辛基召开的奥运会提供服务（此届奥运会推迟到1952年）。然而，无论在技术手段还是建筑工艺上，拉西帕拉特西都是以一种高效和持久的方式建造的，在很大程度上是一座非常超前的建筑。随处可见的霓虹灯和带遮阳板的大窗户也证明了这一点。

卡地亚·萨博拉宁：这座建筑是芬兰悠久建筑传统"完整的艺术作品"中的一员。现如今，阿莫斯·瑞克斯博物馆项目使拉西帕拉特西很好地诠释了对建筑的保护如何与新的当代建筑恰到好处地融合在一起，这一点很有前瞻性。我们希望融合的结果恰如其分，同时也充满活力，令人耳目一新，神清气爽。

采访人：您是如何在这样的背景下进行此项目的？请您讲讲您对拉西帕拉特西的记忆，讲讲您的修复标准以及您选择的施工技术。

卡地亚·萨博拉宁：在1999年，塔利建筑师事务所完成了对拉西帕拉特西的翻新工作。当时，目标是通过复制等修葺技术恢复拉西帕拉特西原本的精神面貌。保护和恢复建筑原本的面貌也是JKMM建筑师

Interviewer *Lasipalatsi was originally intended for temporary use. But unexpectedly it has been standing for more than 80 years, even being recognized as one of the Finnish Modernism masterpieces by Docomomo.*

Asmo Jaaksi Lasipalatsi is the work of a young trio of architects: Kokko-Revell-Riihimäki. Their youthful enthusiasm, exceptional skill and heartfelt passion for this project exudes from the building even today. The design is nonetheless very much tied to its time and represents an early and even visionary example of Finnish Functionalism. Itself is also a historically sensitive context as it neighbors the listed 19th century's neo-classical barracks. The listing of the barracks was contingent on the retention of the former military parade ground – now Lasipalatsi Square – as an unbuilt open public space. This has been achieved by building the Amos Rex Museum galleries underground.

Interestingly, Lasipalatsi was initially conceived as a mixed-use, multipurpose pavilion type facility with a restaurant, shops, and a cinema to temporarily serve the 1940 Olympic Games in Helsinki (postponed to 1952). Nonetheless built in an efficient and lasting way, both technically and architecturally, it is very much a building ahead of its time. Its readily visible neon lights and large windows with their sun blinds also attest to this.

Katja Savolainen The building is a part of a long-standing tradition in Finnish architecture of a "complete work of art". It feels forward-looking now in the way the Amos Rex Museum project has enabled Lasipalatsi to showcase an architectonic response to how the conservation of buildings can coincide happily with new contemporary architecture. We hope the result feels appropriate but also dynamic and invigorating.

Interviewer *How did you approach the project with such background? Please share your retrospection of Lasipalatsi, your criteria for the restoration, and construction techniques you chose.*

Katja Savolainen The previous renovation was completed in 1999 by Talli Architects. Its objective was to reinstate the spirit of Lasipalatsi using reconstruction techniques including replicas. Conservation and bringing back the original qualities of the building was also the aim of JKMM Architects' restoration works. The existing building and even the existing replicas from the previous refurbishment have been preserved whenever possible. In the new interventions, we have sought to reach the same standards of craftsmanship as had been achieved in the 1930s. New paint analysis tests and further research unveiled valuable information. For instance, the main rooms have now been made brighter in tone to mirror what they would have been like when the building first opened to the public. Nonetheless, when appropriate, we also wanted some of the building's patina to still show through. It is important to note that we did not seek to replace existing restoration works but instead accepted these as a part of the story and layering of the building's history.

As a point of methodology, we wanted to ensure that when Amos Rex opens, the 1930s building can continue to function just as it had before – in addition to its new role support-

事务所修复工程的目标，尽可能保留原有建筑，甚至以前翻新的复制品。在新增建的项目上，我们力求达到与20世纪30年代相同的工艺标准。新的油漆分析测试和进一步的研究揭示的信息很有价值。例如，主房间现在修复得更加明亮，基本再现了建筑物首次向公众开放时的情景。尽管如此，在适当的情况下，我们还是希望建筑的一些古铜色或锈迹斑斑仍然能够展现出来。值得注意的是，我们并没有试图取代原来的修复工程，而是接受这些，将其看作是该建筑历史故事和层次的一部分。作为一种方法论，我们希望确保当阿莫斯·瑞克斯博物馆开放时，除了支持博物馆运营的新角色，其中20世纪30年代的建筑功能也能像以前一样继续发挥作用。该项目确实保证了在不减弱建筑本身历史意义的同时，使建筑焕发新的生机。

采访人：要创造这样一个既有趣又令人兴奋的新空间，您必须决定需要修复多少旧的，又需要增加多少新的。为实现这一目标，您专门选择了怎样的设计理念和空间策略?

阿斯莫·亚克思：我们必须谨慎考虑，如何在20世纪30年代的建筑背景和适应当前的需求之间取得平衡。事实上，我认为这是此项目最具挑战的方面之一。

拉西帕拉特西与阿莫斯·瑞克斯博物馆新馆的衔接区域，由JKMM建筑师事务所设计。当然，这样设计的目的性很强，我们希望它也具有实验性的突破，目的是利用原有的建筑氛围在新展馆空间中发挥其作用。

参观者通过建于20世纪30年代的拜尔·瑞克斯电影院的大门进入建筑，到达拉西帕拉特西的门厅，通过倾斜向下的楼梯走进展厅。新馆的屋顶由一系列圆顶组成。透过屋顶的天窗，人们可看到外面建筑物的一角。策展人精心摆放展品，可使自然光透过天窗投射到展品上。即使在地下6m处，参观者仍可以体验置身于地面的感觉，就像身在城市某个特定区域一样。

天花板的圆顶以及背后所蕴藏的历史内涵，使整个展览大厅的规模给参观者带来敬畏感，营造出一种戏剧般的氛围。顶部采光的展厅内光线微妙，有着如同天窗照明一般的效果。

采访人：您提议通过将展览空间移至地下，释放地面，使其成为一个新的公共广场，从而协调新旧建筑，增加公共空间的价值。

阿斯莫·亚克思：拉西帕拉特西作为芬兰20世纪30年代首屈一指的建筑，融合在阿莫斯·瑞克斯博物馆项目之中，称得上是种动态的体验。大胆添加新的部分，会带来一种过去与现在相结合的体验。我们希望这样的设计除了是一种无缝衔接，也给博物馆空间带来时代气息。新旧元素之间的对话是最明显的特征。在这样的大规模建筑中，人们可以看到不同时期的芬兰建筑。其中，拉西帕拉特西的外部与博物馆那雕塑般的屋顶空间更加紧密地结合在一起。毗邻拉西帕拉特西的广场是赫尔辛基最重要的公共空间之一。我们希望新建成的广场及其和缓起伏的圆顶能够作为赫尔辛基城市文化的一部分而为大家所接受，希望城市中的每一个人都有归属感。

ing the Museum's operations. It really was about bringing the building back to life without compromising the things that make it historically less significant.

Interviewer *You had to decide how much of the old to maintain and how much of the new to add, yet to create such an intriguing and provoking new space. What kind of architectural device and spatial strategy did you specifically implement to achieve the goal?*

Asmo Jaaksi We had to strike a balance with how current requirements can be adapted sensitively in a 1930s' context. I think this was, in fact, one of the most challenging aspects of the project.

The areas in the project where Lasipalatsi meets the architecture of the museum's new gallery spaces have been designed by JKMM so that this is, of course, purposeful but, we hope, also experimental. The intention has been to work with the atmosphere of the pre-existing building and to play on it in the gallery spaces.

Visitors enter the building through the 1930s' doors of the Bio Rex cinema, reach Lasipalatsi's foyer, and walk down to the exhibition hall via sloping stairs. The roof of the new gallery is formed by a series of domes with angled roof lights that frame views of the surrounding buildings and allow exhibitions to be lit with natural light if the curators choose. Thus, visitors can experience a sense of place and feel located in a specific part of the city even when they are six meters below ground.

The hall's scale is awe-inspiring with the ceiling domes, and their historical associations, creating their own particular sense of drama. The subtle quality of light in the top-lit exhibition hall is similar to that achieved through clerestory lighting.

Interviewer *You proposed to free up the ground level as a new public square by bringing down the exhibition spaces underground, consequently harmonizing the old and the new and adding value as public space.*

Asmo Jaaksi Integrating one of Finland's architecturally pioneering 1930s' buildings – Lasipalatsi – as part of the Amos Rex project has been a moving experience. By adding a bold new layer to Lasipalatsi, we feel we are connecting the past with the present. We would like this to come across as a seamless extension as well as an exciting museum space very much of its time.

The dialogue between new and old elements is most visible where one can see different periods of Finnish architecture coming together at a larger scale, including the more intimate coming together of Lasipalatsi's exteriors with Amos Rex's sculptural roof space.

The square adjacent to Lasipalatsi is one of the most important public spaces in Helsinki. We hope the newly landscaped Lasipalatsi Square with its gently curving domes will be received as a welcome addition to Helsinki's urban culture; a place everyone and anyone in the city can feel is their own.

皇家剧场

Royal Arena

3XN + HKS

剧场能否算得上芳邻？大型体育场馆和音乐场馆往往位于市郊，但在哥本哈根却并非如此。刚刚开放的皇家剧场位于一片密集住宅区的中心地带，面积为35 000m²，是专为音乐会和国际体育赛事而设计的场馆。

皇家剧场是哥本哈根最受期待的文化场所之一。它肩负着两项重任：一方面是打造一个富有吸引力、灵活性强的多功能舞台，既能招揽本地观众，也能吸引国际观众；另一方面是确保该建筑本身能促进整个地区乃至城市的发展。

芳邻

毫无疑问，如此规模的建筑会影响与之相邻的社区。针对竞技场类型的建筑设计，必然会有"以人为本"的彻底反思。若想成为"芳邻"，关键是要对该地区做出艺术贡献，与所在社区建立亲密的互利互惠合作关系，让周围的环境更具活力，并为在其附近生活和工作的人提供新的机遇。在设计方面，建筑的极简北欧风格，建筑外层设计巧妙、呈曲线形排列的木质肋片，与其周围区域相得益彰。

斯堪的纳维亚风格的剧场

皇家剧场设计的核心是一座独树一帜的裙楼，它是连接周边的纽带。在裙楼的外缘上设计了若干小型广场、营业场所和活动区，并且能有效地从这些区域吸引观众流，这便是裙楼设计的宗旨所在，同时，它也鼓励社区包容各种公共空间、阶梯和毗邻区域——在剧场空闲时，这些区域也能提升社区的活力。游客可通过宽敞的楼梯进入裙楼，然后再经由一个大型主入口进入剧场；如果观众人数众多，也可以选择沿着建筑立面均匀分布的四个入口进入。在剧场的正常入口上方，由肋片组成的波浪微微抬高，使游客寻找路线更加方便和有规律。

从金属乐队到席琳·迪翁——从冰上曲棍球到太阳马戏团

碗状的圆形剧场具有极高的灵活度，剧场通过倾斜的墙壁、平坦的天花板、隔声墙、出入口和一流的舞台设置等方式来竭力提升表演质量。此外，对称分区布局能够以最恰当的方式为每场音乐会新增、减少、扩展或分割座席，还能够根据门票销量进行快速的运营策略变更，满足需求。舞台周围的座位有22m高，能保证观众无论坐在哪里，都看得到焦点。大多数观众坐在舞台、跑道、球场的三面，因此可选择在第四面容纳更多游客，举办体育赛事和特殊文化活动。在音乐会模式下，竞技场地板可以收起。高度灵活的设计让场馆可以容纳各类赛事、表演，因此几乎可以进行无限的类型配置。

皇家剧场设有裙楼，为社交会议和日常活动提供各种公共区域；温暖的木材立面让观众可以面向室外展望，好奇者也可以向室内观望。皇家剧场能够开启都市生活，提升附加价值，并融入周边环境。

Can an arena be a good neighbor? Large stadiums and music venues are often placed in the outskirts of cities, but not in Copenhagen. Royal Arena, a 35,000m² venue specially designed for concerts and international sporting events, has just opened in the middle of a dense residential area.

As one of the most anticipated cultural venues in Copenhagen, it combines two key ambitions: to create an attractive and highly flexible multi-purpose arena that can attract spectators locally as well as internationally, while ensuring the building's presence to catalyze growth for the entire district as well as the city.

ROYAL AR
A
ROYAL ARENA

NA
field's

©Rasmus Daniel Taun (courtesy of the architect)

The Good Neighbor

Without a doubt, a building of this size affects the community next to it. A radical rethinking of the Arena typology, "putting people first", was inevitable. To become a "good neighbor", it was crucial to create an aesthetic contribution to the area and an intimate symbiosis between the building and the community, activating its surroundings and offering new opportunities for those who live and work adjacent to the building. Design-wise, its curvy wooden fins and the minimalistic Nordic expression fit the nearby area.

A Scandinavian Take on the Typology

Central to the design of the Royal Arena is a unique podium acting as a link to the adjoining neighborhood. A single podium is designed to efficiently absorb the movement of spectators through a variety of small plazas, pockets and gathering areas which have been carved from the podium's perimeter. It simultaneously encourages the community to embrace the variety of public spaces, staircase, and adjacencies which promote activity and liveliness when the building is not in use. Visitors enter the podium via a wide staircase and from the podium enter the building via a large main entrance or, in case of large audience sizes, are distributed smoothly along the facade between four different entries. The wavelike movements lifted slightly above the natural entry points of the Arena make way-finding easy and logical.

From Metalica to Celine Dion – From Ice Hockey to Cirque du Soleil

This extremely flexible bowl endeavours to improve performance experience through angled walls, a flat ceiling, acoustic walls, vomitories, and a first-rate stage set up. In addition, the symmetrical block layout, allows seating to be built up, reduced, expanded or sectioned off in the most appropriate ways for each concert, but also for quick operational changes and requirements based on ticket sales. With a 22-meter height around the stage, it ensures the focal point no matter where one is seated. Most of the spectators are seated on three sides of the stage/track/court, with the option to accommodate further visitors on the fourth side for sporting events and special cultural events. In concert-mode, the arena floor can be retracted. The flexibility of the design allows for the widest range of events, and possible configurations are therefore almost infinite.

With a podium that offers different public areas for social meetings and daily activities and a warm timber facade allowing spectators to look out and the curious to look in, Royal Arena triggers urban life, adding value and fitting into the surrounding neighborhood.

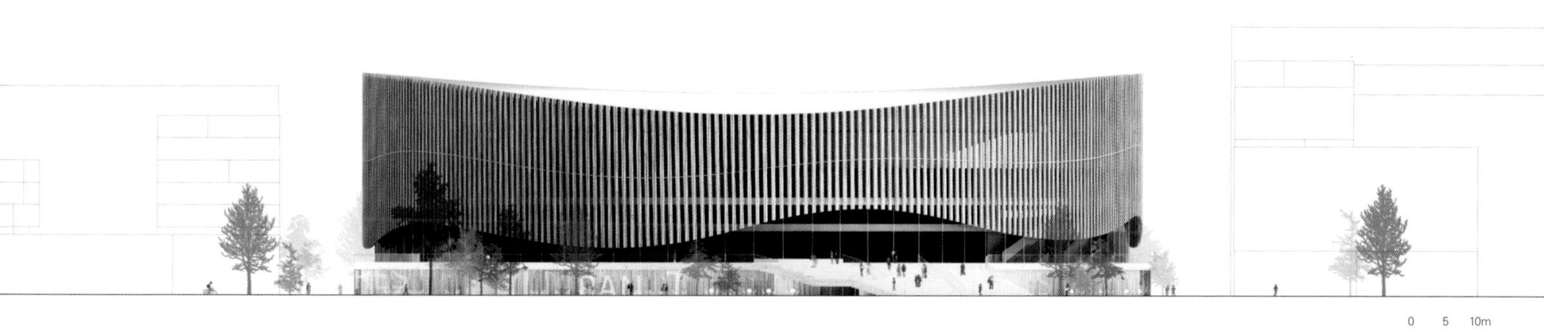

南立面 south elevation

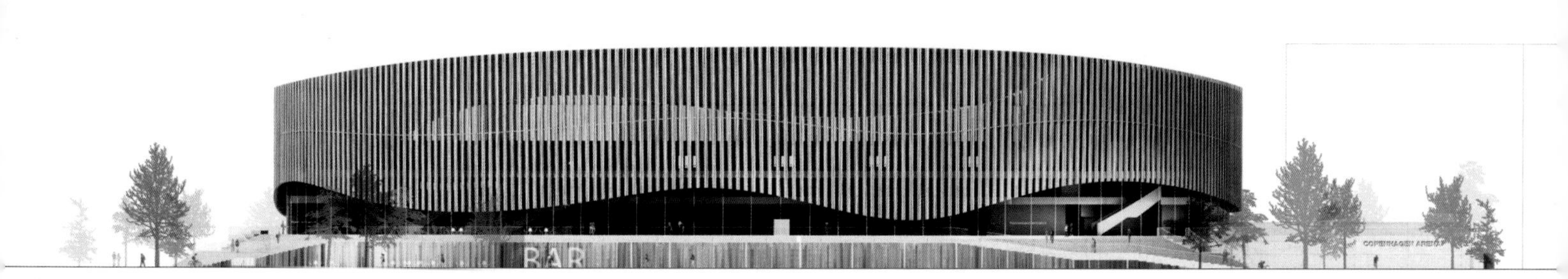

东立面 east elevation

VIP 113-115

310-318
112-118
109-11

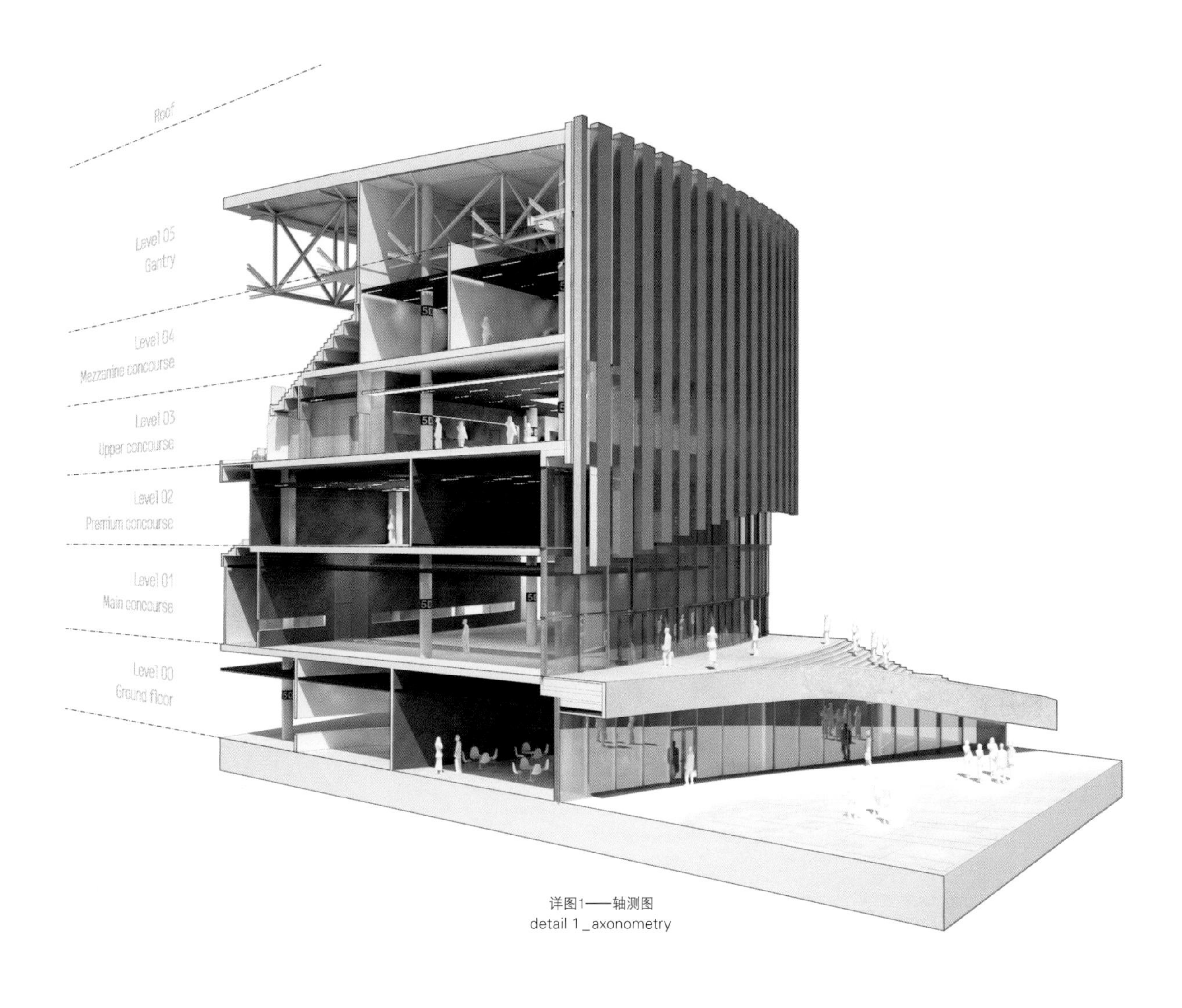

详图1——轴测图
detail 1_axonometry

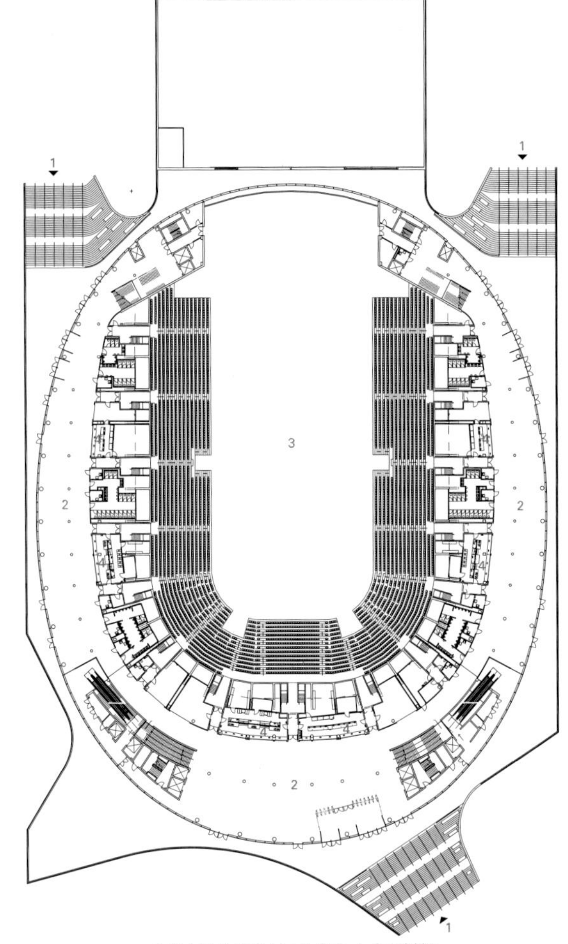

1. 主入口 2. 中央大厅 3. 舞台 4. 食品饮料区
1. main entrance 2. concourse 3. stage 4. food and beverages
二层 first floor

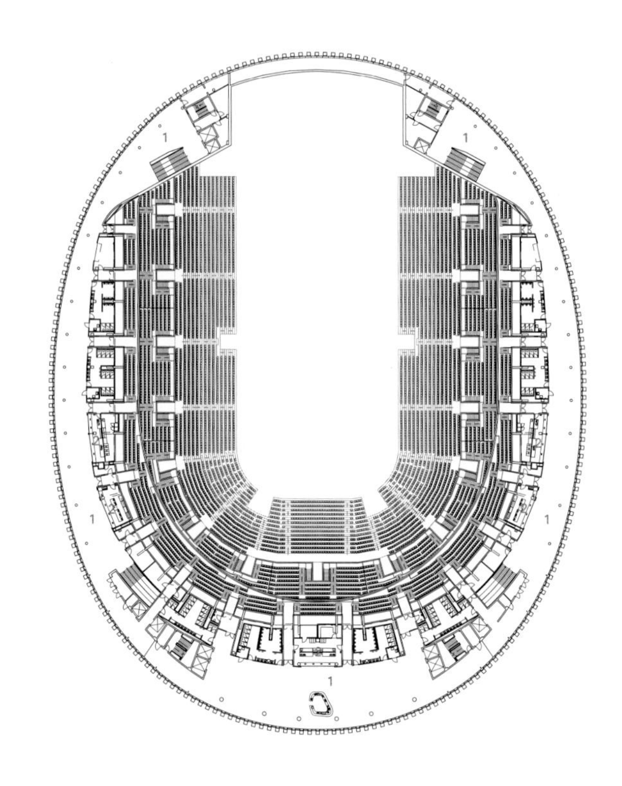

1. 中央大厅 2. 食品饮料区
1. concourse 2. food and beverages
四层 third floor

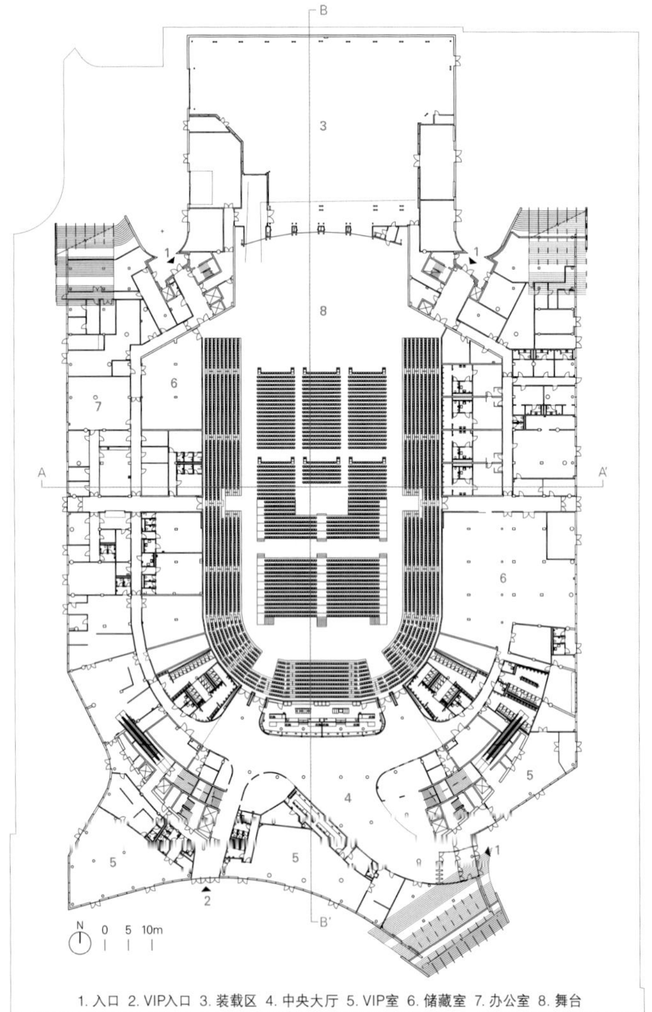

1. 入口 2. VIP入口 3. 装载区 4. 中央大厅 5. VIP室 6. 储藏室 7. 办公室 8. 舞台
1. entrance 2. VIP entrance 3. loading bay 4. concourse 5. VIP room 6. storage 7. offices 8. stage
一层 ground floor

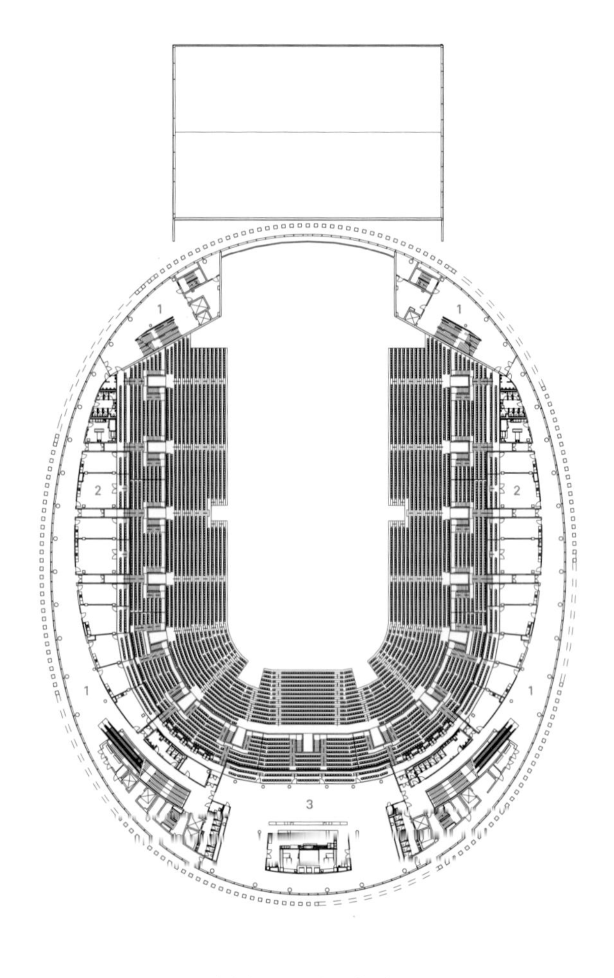

1. 中央大厅 2. VIP室 3. 餐厅与酒吧
1. concourse 2. VIP room 3. restaurant and bar
三层 second floor

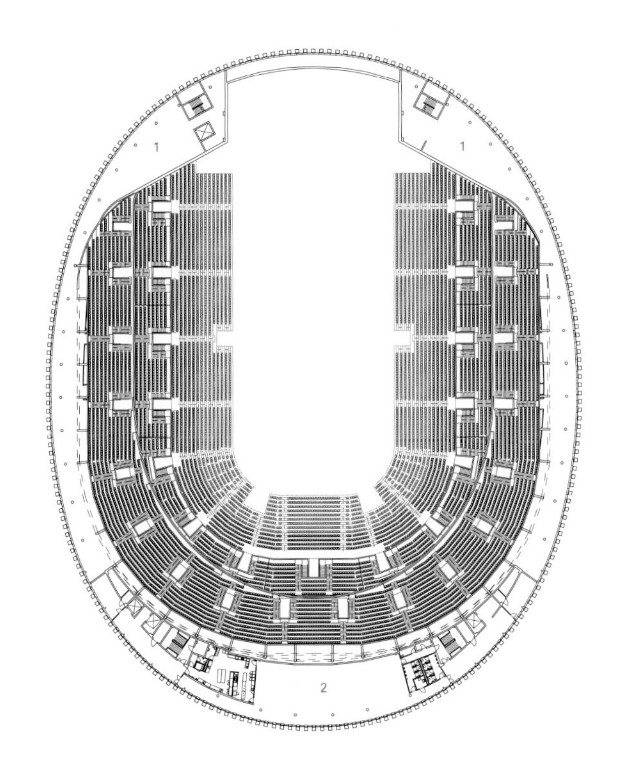

1. 中央大厅 2. 餐厅与酒吧
1. concourse 2. restaurant and bar
五层 fourth floor

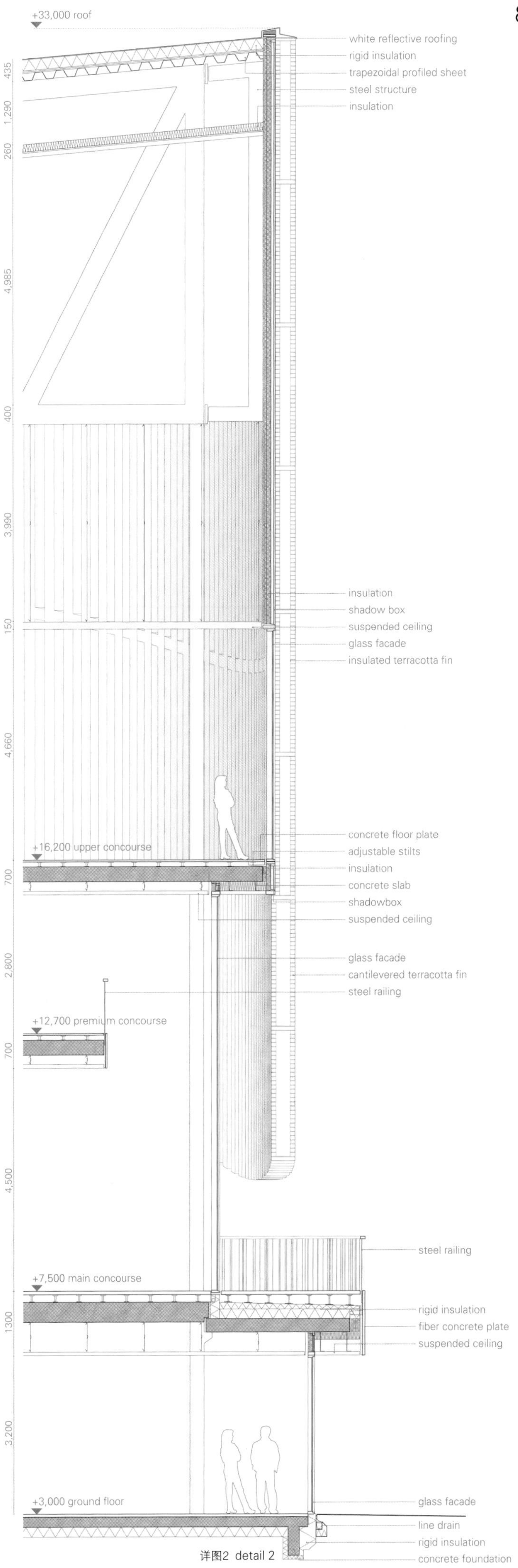

详图2 detail 2

项目名称：Royal Arena / 地点：Copenhagen, Denmark / 事务所：3XN (lead consultant and design architect), HKS (arena specialist) / 项目团队：Kim Herforth Nielsen, Jan Ammundsen, Bo Boje Larsen, Peter Feltendal, Audun Opdal, Maria Tkacova, Jack Renteria, Robin Vind Christiansen, Dennis Carlsson, Andreas Herborg, Anja Pedersen, Bodil Nordstrøm, Christian Harald Hommelhoff Brink, Gry Kjær, Ida Schøning Greisen, Jakob Wojcik, Jan Park Sørensen, Jeanette Hansen, Juras Lasovsky, Laila Fyhn Feldthaus, Mads Mathias Pedersen, Marie Persson, Mikkel Vintersborg, Pernille Ulvig Sangvin, Sang Yeun Lee, Sebastian le Dantec Reinhardt, Simon Hartmann-Petersen, Stine de Bang, Sune Mogensen, Søren Nersting, Tobias Gagner, Torsten Wang, Henrik Rømer Kania / 工程师：Arup, HAMI, ME Engineers / 景观建筑师：Planit-IE / 客户：Arena CPHX P/S / 用途：Multi-purpose Arena for international concerts, culture and sports events / 建筑面积：37,000m² / 建筑规模：35m with the capacity of 16,000 people and 12,500 seatings / 竣工时间：2017 / 摄影师：©Adam Mørk (courtesy of the architect) (except as noted)

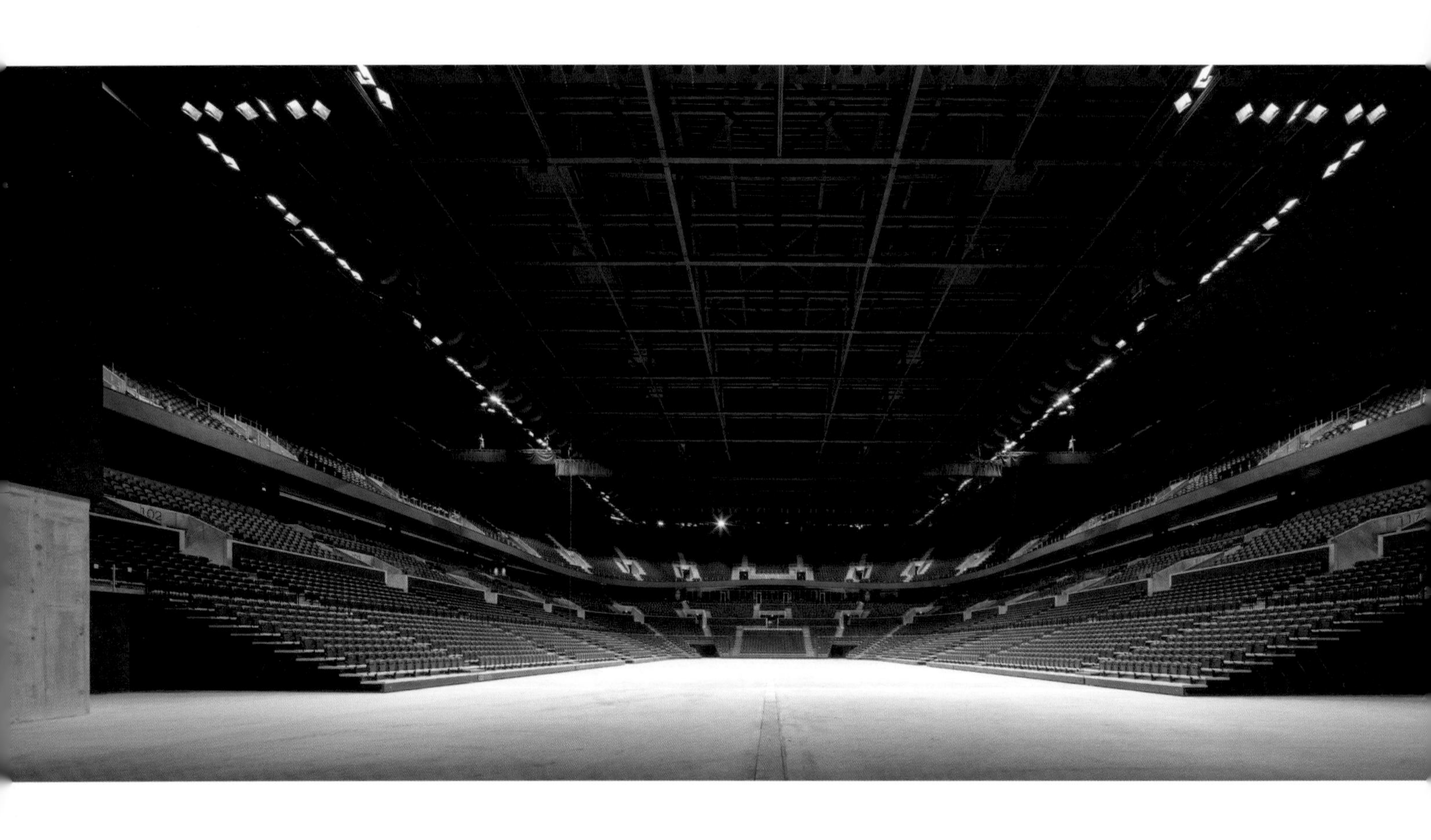

114
113

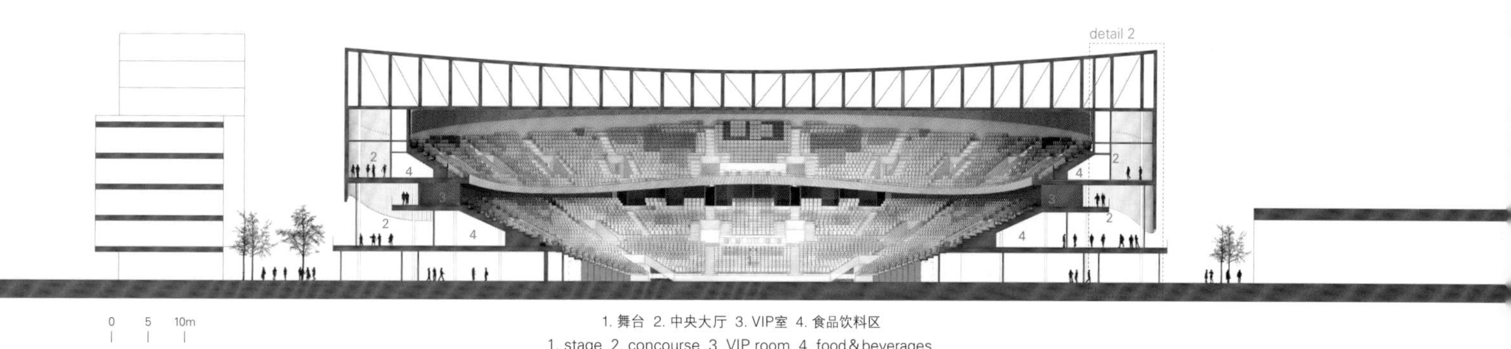

1. 舞台 2. 中央大厅 3. VIP室 4. 食品饮料区
1. stage 2. concourse 3. VIP room 4. food & beverages
A-A' 剖面图 section A-A'

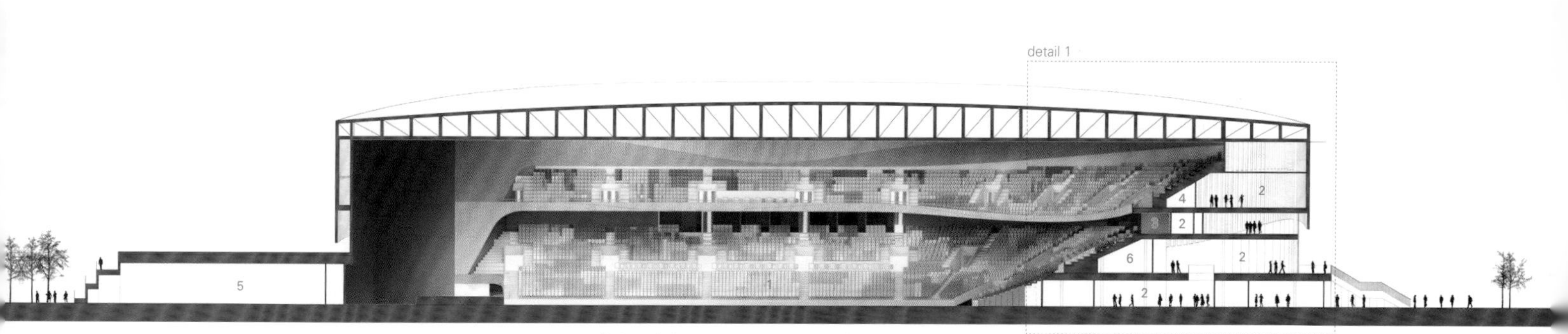

1. 舞台 2. 中央大厅 3. VIP室 4. 食品饮料区 5. 装载区 6. 卫生间
1. stage 2. concourse 3. VIP room 4. food & beverages 5. loading bay 6. toilets
B-B' 剖面图 section B-B'

巴勒斯坦博物馆
The Palestinian Museum

Heneghan Peng Architects

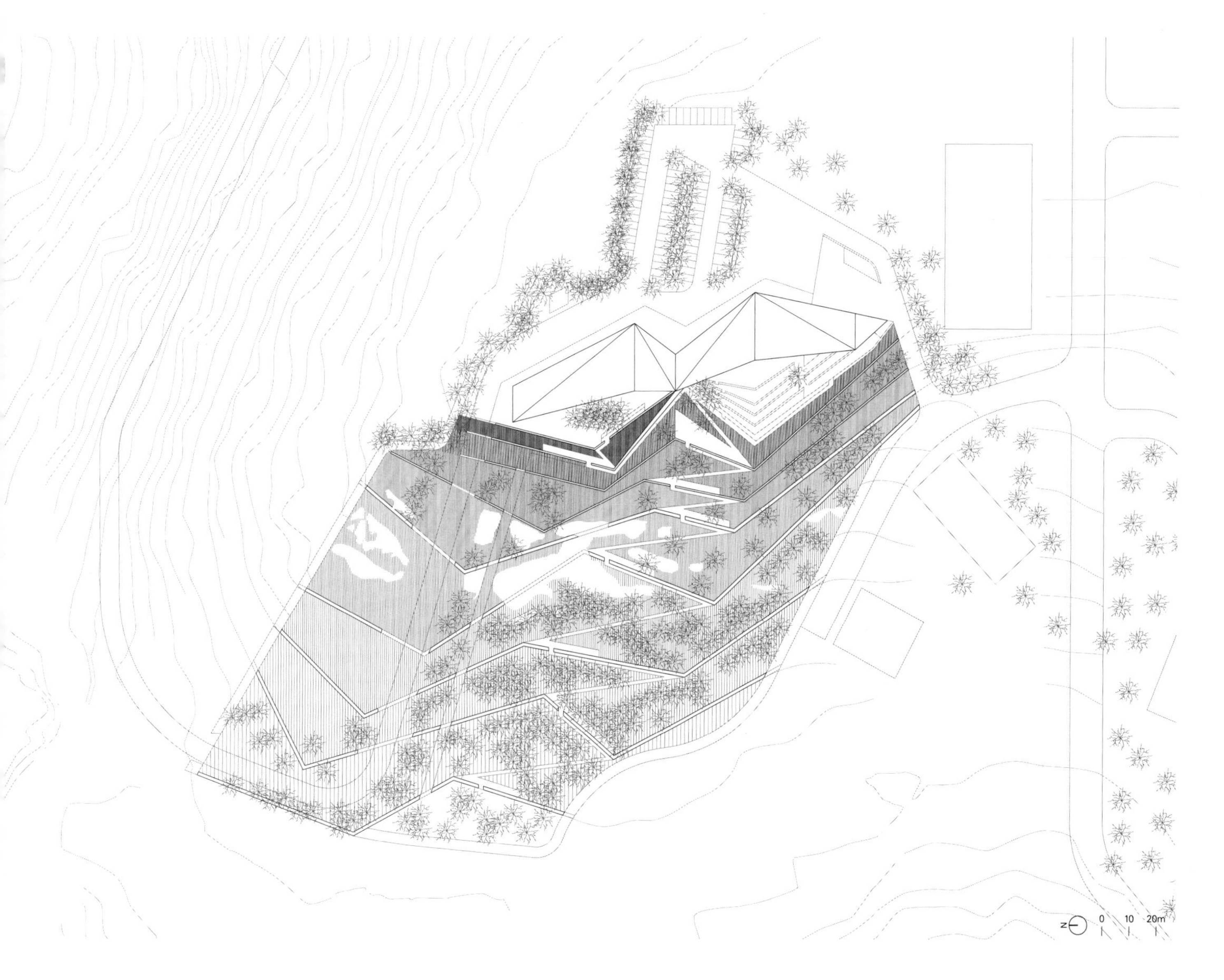
N
0
10
20m

打造一个塑造和传播巴勒斯坦历史、社会和文化知识的平台，使平台遥遥领先，可信度最高，稳健性最强，这是巴勒斯坦博物馆的使命。为达成这项使命，客户有意在耶路撒冷（西岸）以北25km处建设巴勒斯坦博物馆中心，分两期进行建筑施工。

第一期工程已于2016年竣工，建筑面积为3500m²，包括温度、湿度可控的展览空间，一座露天剧场，一间带有户外座位的自助餐厅，一间图书馆，多间教室和储藏室，一家礼品店以及办公场所。以上所有场所都规划在花园内的4ha面积中。在工程第二期，博物馆总面积将扩张到10 000m²。

巴勒斯坦的景观当中能体现出对城市的用心经营，每一个元素都经过了精心设计，设计师利用它们去讲述国家干预、生产、文化、环境和商业的历史。与这座城市一样，梯田景观也融入了历史。

巴勒斯坦博物馆的设计方法，便是讲述梯田景观的历史，博物馆嵌在梯田场地之间，利用这片场地去更好地阐述多元文化。

该场地由一系列层叠的梯田组成，这些梯田由田野石墙建造，沿用了该地区以前的农业梯田的设计理念。

从文化景观到原生态景观，景观主题在整个梯田上展开。靠近博物馆的梯田都经过人工精心设计，植物经过精心栽培，人们沿着梯田向西走，就会发现所栽种的植物逐渐发生变化。

层叠的梯田种植了不同的植物：通过贸易路线引入的柑橘，原生香草，丰富多样的景观。梯田主题包括：文化景观、与文化和历史相关的主题；农业遗产；植物与贸易路线和商业的关系；与荒野和本土植物、灌木丛生地区、草地有关的自然景观和主题；自然与文化——将原生态植物纳入本地农业和食品药品。建筑本身与景观融为一体，在山顶勾勒出鲜明的轮廓，既融入了景观，又创造了一种具有高辨识度的坚毅形象。博物馆大部分是单层建筑，由南向北顺着山坡依势而建，俯瞰西面的花园。一层包括：入口接待处，博物馆管理处，展厅，放映室，以及位于最北面、通向花园的咖啡厅，和俯瞰着南端的一个由石头铺设的露天剧场。再往下一层，有公共教育和研究中心，包括教室、工作室和行政办公空间。教育中心向西通向由石头铺设的露天剧场。

除了教育和研究中心外，主要的艺术收藏空间、摄影档案室和艺术管理室也都在一层的下层。这些区域不对公众开放，它们通向建筑东侧的安全交货点。该建筑将成为巴勒斯坦第一座获得LEED（能源与环境设计）认证的建筑。

The mission of the Palestinian Museum was to be the leading, most credible and robust platform for shaping and communicating knowledge about Palestinian history, society and culture. To deliver on this mission, the client's intention has been to develop the Palestinian Museum hub located 25km north of Jerusalem (West Bank), in two phases of building construction.

Phase 1 (completed in 2016) consists of a built area of 3,500m². It includes a climate-controlled gallery space, an amphitheater, a cafeteria with outdoor seating, a library, classrooms, storage, a gift shop and administrative spaces; all set within 4 hectares of planned gardens. During Phase 2, the Museum will expand to a total of 10,000m².

The landscape of the Palestine has the "worked" quality of a city; every element of it has been touched and tells a story of intervention, production, culture, environment and commerce. Like a city, the terraced landscape has embedded within its history.

The approach to the Palestinian Museum was to draw on this history of the terraced landscape, embedding the museum into its immediate site and guide this site to tell a larger story of a diverse culture.

The site is formed through a series of cascading terraces, created by field stone walls which trace the previous agricultural terraces of the area.

The theme of the landscape, from the cultural to the native landscape, unfolds across the terraces with the more cultured and domesticated terraces close to the building, the planting changes gradually as one moves down the terraces to the west.

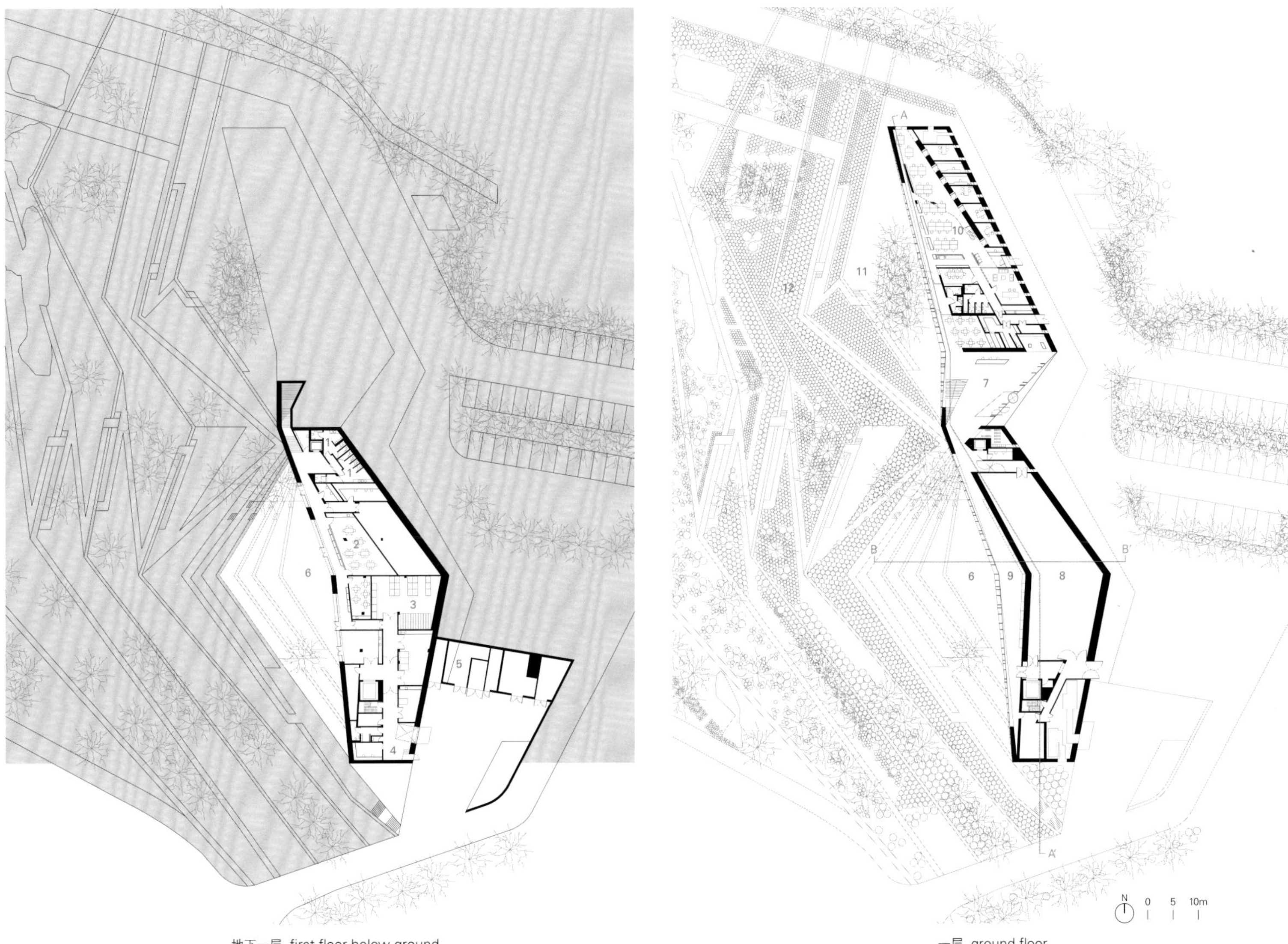

地下一层 first floor below ground

一层 ground floor

1. 卫生间 2. 教室 3. 储藏室 4. 交货点 5. 工厂 6. 下沉露天剧场
7. 入口 8. 气候可控画廊 9. 不可控画廊 10. 行政办公空间 11. 咖啡露台 12. 博物馆花园
1. wc 2. classroom 3. storage 4. delivery 5. plant 6. sunken amphitheater 7. entrance
8. climate controlled gallery 9. non-controlled gallery 10. administrative spaces 11. cafe terrace 12. museum gardens

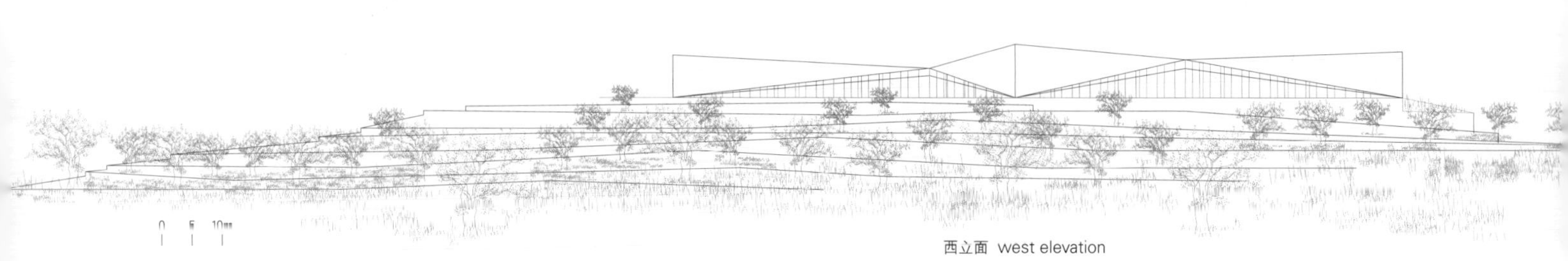

西立面 west elevation

AERO
POS
87
TALE

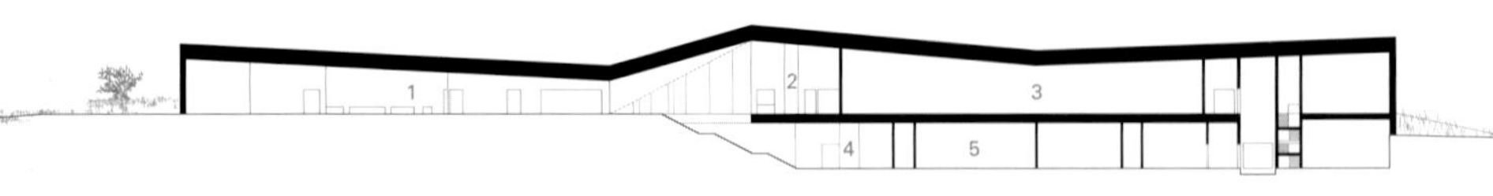

1. 行政办公空间 2. 入口 3. 气候可控画廊 4. 卫生间 5. 教室
1. administrative spaces 2. entrance 3. climate controlled gallery 4. wc 5. classroom
A-A' 剖面图 section A-A'

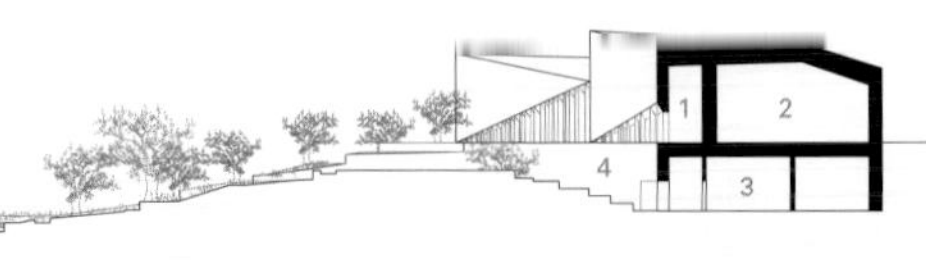

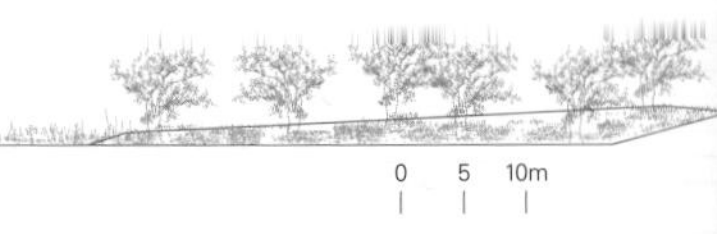

1. 不可控画廊 2. 气候可控画廊 3. 教室 4. 下沉露天剧场
1. non-controlled gallery 2. climate controlled gallery 3. classroom 4. sunken amphitheater
B-B' 剖面图 section B-B'

The cascade of terraces tells diverse stories; citrus brought in through trade routes, native aromatic herbs, a rich and varied landscape. Terrace themes include: Cultural Landscapes and themes relating to culture and history; Agricultural Heritage; Relationship of plants to trade routes and commerce; Natural Landscapes and themes relating to wilderness and native plants, scrub lands, grass lands; Nature & Culture. The aim is to incorporate native plants into domesticated agriculture and food/medicine.

The building itself emerges from the landscape to create a strong profile for the hilltop both integrated into the landscape yet creating an assertive form that has a distinctive identity. Largely in single-storey, it stretches out along the hilltop from the south to north, overlooking the gardens to the west. The ground floor, comprising entrance reception, museum administration, galleries, screening room and cafe opens out directly to the gardens at its northern end, while overlooking a stone amphiteater below it at the southern end. On the lower ground floor, there is a public Education and Research Center with classrooms, workshops and administrative spaces. The education center opens out to a cut stone amphiteater to the west.

In addition to the Education and Research Center, the main art collections spaces, photographic archives, and art handling are all located on the lower ground floor. These spaces are not accessible to the public; they open out to a secure delivery yard at the eastern side of the building. The building will be the first LEED Certified building in Palestine.

项目名称：The Palestinian Museum / 地点：Birzeit, Palestine (West Bank) / 事务所：Heneghan Peng Architects / 项目经理：Projacs International / 机电管道、防火：ARUP-concept, scheme design; Arabtech Jardaneh-tender, construction stage / 工料测量师：Davis Langdon/AECOM-concept, scheme design; Arabtech Jardaneh-tender, construction stage Concept facade: T/E/S/S / 概念立面：Bartenbach GmbH / 概念照明：Lara Zureikat / 景观建筑师：Consolidated Contractors Company, Tubeileh, Al Sabe Landscape / 承包商/客户：Welfare Association-Taawon / 用地面积：39,000m² / 建筑面积：3,500m² (Phase 1) + 40,000m² Gardens / 造价：Euro 16.5m / 施工周期：24 months 竣工时间：2016 (Phase 1) / 材料：Dolomitic Limestone-Nassar Stone Ltd., Bethlehem, Palestine; Schueco Curtain Walling & bespoke brise soleil-ALICO, Dubai, UAE; Tretford Carpet-Waterford Carpets Ltd., Ireland; ERCO Lighting-ERCO, Germany; D-Line Ironmongery-D-Line, Dubai, UAE; Bespoke Fitted furniture-Manjorco, Ramallah Palestine; Signage, wayfinding-Creative Ad Design, Beit Jala, Palestine / 摄影师：©Iwan Baan (courtesy of the architect)

共享空间

Space for Eve

现如今，不断变化的数字化技术刺激并已经彻底改变了人们的生活。在应对这样的变化时，我们对空间充满了迫切需求，这样的需求可以使人们从根本上感到满足。从这一意义来看，建筑便重新获得了其基本价值，即成为现实世界与人类个体世界进行邂逅与物质交换的场所。

这部分收集的项目讲述了在公共场所安置居民的可能性策略，即使它们都是私有空间，因此我们可以注意到人们对空间管理的极致研究，它可以将集体的外部空间转换为大型的室内场所。室内设计是不属于外部

Nowadays we are witnessing a pressing demand for spaces in which people can meet physically, as a need to counteract the continuous digital stimulus which has radically changed our lives. In this sense, architecture regains its foundational value: being the place of encounter and physical interaction between the real world and the individual universe of every human being. The projects collected here narrate possible strategies to host the inhabitants within public places, even when they are private spaces: for this reason we can notice an extreme research on space control, which transforms the collective exteriors into large interiors. It can be seen

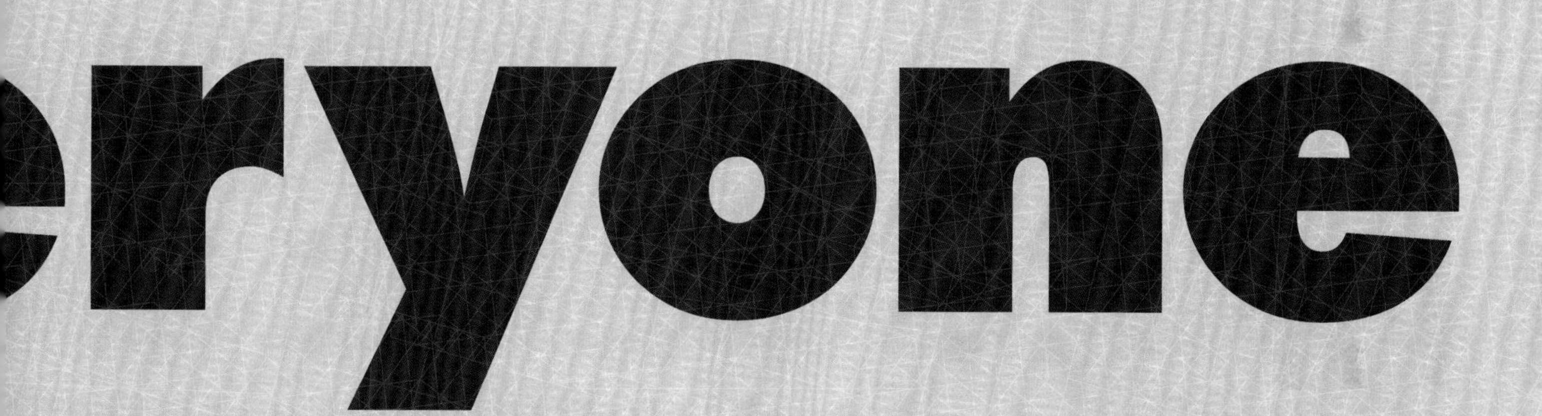

世界的，这不仅可以在细节和材料中看出来，而且还表现在建筑元素之中，如经常设计使用的楼梯、休息区、沉思区以及各种铺地材料等。

这些项目打破了内部与外部空间、公共与私有空间、自然与人造空间的界线。所打造的建筑促进了人类活动，将社区居民聚集在一起，并恢复了外部空间的社交用途。

from details and materials, which do not belong to the world of the exteriors, but also from the elements of architecture, with frequent use of stairways, rest and contemplation areas, different pavements.

They are projects that destroy the limit between internal and external spaces, between public and private spaces, between nature and artifice: they provide architecture with the task of fostering human actions, defining functions that can summon the community, returning to a social use of the external world.

共享空间
Space is for Everyone

Diego Terna

自由空间是一个充满机会的空间，一个民主的空间，一个尚未构思用途的空间。无论是否有意，人与建筑都会相互交流，即使是在建筑师离开很久之后，建筑本身也会找到与人共享和共存的方式。

（……）我们相信人人有权从建筑中受益。建筑本身的意义是为我们提供庇护，改善心情。即使未进入建筑物，行人看到路边精美的墙壁也会心生愉悦。

通过拱门瞥见庭院，或倚靠在阴凉处或遮风避雨的隐蔽处，也是如此。

——伊冯·法雷尔和雪莱·麦克纳马拉，自由空间，第16届威尼斯国际建筑展

上次威尼斯国际建筑展期间，策展人伊冯·法雷尔和雪莱·麦克纳马拉提议思考一个主题，这个主题看似相当笼统，不是十分具体，但却可以引起人们对公共空间哲学的深深思考。他们断言每座建筑都与公共环境有关，就连与房屋相连的墙壁也包括在内，因为精美的墙壁可以提升每个人所在的空间的质量。两位爱尔兰建筑师给每座建筑的周边环境赋予了社会意义。

事实上，建筑一直扮演着重要的角色，它不仅仅是严格履行项目所产生的功能，对于游牧民族而言，建筑还为他们提供了起到临时保

FREESPACE can be a space for opportunity, a democratic space, un-programmed and free for uses not yet conceived. There is an exchange between people and buildings that happens, even if not intended or designed, so buildings themselves find ways of sharing and engaging with people over time, long after the architect has left the scene.

[…] We believe that everyone has the right to benefit from architecture. The role of architecture is to give shelter to our bodies and to lift our spirits. A beautiful wall forming a street edge gives pleasure to the passers-by, even if they never go inside. So too does a glimpse into a courtyard through an archway, or a place to lean against in the shade or a recess which offers protection from the wind and rain.

– Yvonne Farrell and Shelley McNamara, *FREESPACE*, 16th International Architecture Exhibition, Venice

During the last International Architecture Exhibition in Venice, the curators Yvonne Farrell and Shelley McNamara proposed to think on a theme that was apparently very general and not very specific, but which could have led to very sophisticated reflections on the theme of public spaces: they assert that every architecture relates itself with a public surrounding, even a wall bordering a property, whose beauty can increase the quality of *the space of everyone*; the two Irish architects have loaded the environment, where every building operates, with social meanings.

In fact, architecture has always played an important role that goes beyond the strict fulfillment of the functions for which a project was born; even in the case of nomadic populations, for whom architecture was a mechanism of light and temporary protection, the ambient dedicated to the meeting, *the place of everybody*, served as a spatial epicenter

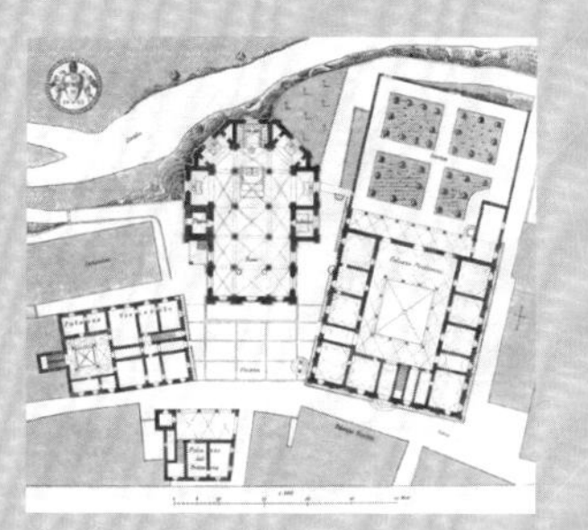

皮恩扎项目Pio II广场平面图
floor plan of piazza Pio II in Pienza

© Oschirmer

皮恩扎中心Pio II，托斯卡纳区，意大利
Piazza Pio II in the center of Pienza, Tuscany, Italy

护作用的轻型结构场所，营造用来集会的专门环境，人人可以共享，可以作为整个社区的空间中心。正如雷纳·班汉姆在《掌控环境的建筑》(1969年)中所说："未建立实体结构的社会，倾向于将他们的活动集中在一些中心焦点上——一个水坑，一片树荫，一堆明火，一位伟大的教师——并居住在外部边界模糊、依功能需求而调节的、几乎没有规律的空间中。"

建筑在履行其社会职能的同时也在慢慢消逝（一堆火、一棵树足以维持需要），却留下了无形却关键的空间痕迹，它可以将人们聚集在一起。这种情况多出现在建筑本身为社区居民创造了实体的共享空间之时。

皮恩扎项目建于14世纪中期，由贝尔纳多·罗塞利诺设计建造，社区中心位于小镇的中央广场，是居民聚集的场所。它就像是一个大型的室内建筑项目，周围环绕着公共与私人建筑：大主教宫、大教堂、市政大楼。这些建筑的立面朝外，给予广场以空间感，细致雕琢的墙壁似乎限定了大厅的界限，大厅面向周围的山谷，引景观入城市，使得人、自然、建筑实现了和谐统一。

正如双年展的策展人所假设的那样，建筑的边缘或界限总是一个公开的事实：它是与全体居民交流的窗口，是用形状、材料、孔洞、光线和阴影书写的指南。从这个意义上讲，很难将建筑定义为纯粹的私有空间。实际上，任何建筑场所都会在某一刻，由私有空间转化为公共场所。

for the whole community, as Reyner Banham observed in *The Architecture of Well-tempered Environment (1969)*, "societies who do not build substantial structure tend to group their activities around some central focus – a water hole, a shading tree, a fire, a great teacher – and inhabit a space whose external boundaries are vague, adjustable according to functional need, and rarely regular."

Architecture fulfills its social role also disappearing (a fire, a tree, are enough), but leaving immaterial, but significant, traces of space that aggregates people. This happens more when architecture exists and it is present, physically, in creating the sharing space for the community.

Pienza, built in the mid-1400s, by Bernardo Rossellino, finds in the main square of the small town, the community core for the meeting of the inhabitants. It is built as a large interior, surrounded by the facades of public and private buildings: the archiepiscopal palace, the cathedral, the municipal building. These facades, facing outwards, create a spatial sense to the public square, thanks to a detailed work on the walls, which seem to define the limit of a large hall, open towards the surrounding valley, allowing the landscape to enter in the city, unifying people, nature, architecture.

The edge of architecture, its limit, is always a public fact, as hypothesized by the curators of the Biennale: it is the interface of dialogue with the population, a sort of manual written through shapes, materials, holes, lights and shadows. In this sense it is difficult to define an architecture as purely private, because, indeed, every built place finds in the passage between the buildings a moment in which the private becomes public.

Architecture then becomes an almost political fact, a community aggregator that can solve (or help to solve) social

©Nigel Young

苹果自由广场，意大利
Apple Piazza Liberty, Italy

那时，建筑几乎就会成为一个政治事实，一个可以通过建筑本身来解决（或帮助解决）社会问题的社区聚合器。

我们正处于这样一个时代——人与人面对面的交流变得越来越重要，这也许是由互联网社交网络导致人们见面和互动日趋减少所造成的。似乎持续的虚拟接触需要在极致的物理空间中找到对应点，其中视觉、触觉、嗅觉和听觉被直观感受所刺激，而不是由数字媒体所介导。人们对建筑社会聚合形式的新兴趣，被越来越多地投入到建筑空间的功能需求以及交流问题的研究中。

例如，我们可以看到，在最近由米兰的福斯特事务所设计的苹果自由广场（44页）中，真正的商店位于地下，在市中心行走的行人是看不见商店的，他们只能看见一个阶梯广场，而广场上的大型喷泉便是入口。广场是私有的，但是有公共通道。该建筑看似简约，实际却非常复杂，它以城市政治为主题（第一点正是私有空间公共化的模糊性），却彰显了建筑如何通过人们之间相遇产生的刺激进行交流。

简而言之，布兰登·科米尔的担忧并未成为现实：我们仍然对罗马1748年的诺利地图留下了深刻印象，该地图描绘了包括教堂内部的公共空间网格，但如今我们的城市规划很少尝试设计公共空间网格以连接公共机构内部，我们很少要求将该公共元素放入建筑设计任务书中。——"城市内部规划"，第33卷：室内设计，2012年。

用法雷尔和麦克纳马拉的话说，下面的项目确实显示了一项令人恼火的研究，即通过需要公共、开放使用的功能来推动体系结构的发

issues through the very act of constructing itself.

We are in a time when the physical aggregation of people is becoming more and more important, perhaps as a sort of common feeling against the ever less physical ways of meeting and interaction between people, through the social network on the Web. It is as if continuous virtual contacts need to find a counterpoint in an extremely physical space, in which sight, touch, smell and hearing are stimulated by direct sensations, not mediated by a digital medium. This renewed interest in the form of social aggregation of architecture increasingly invests the functional demands, but also the communicative issues, of the architectural space.

We can observe, for example, the recent *Apple Piazza Liberty* (p.44), designed by Foster + Partners in Milan: the real store is located underground, invisible for the persons walking in the city center, while the visible space consists of a stepped square, from which a large fountain, that serves as an entrance, emerges. The square is private but with public access. It is a very complex place, in its apparent simplicity, which invests themes of urban politics (the first of which is precisely its ambiguity of private space with public use), but which strongly reveals how the communication of a product passes also through a stimulus of the encounter between people in architecture.

In short, the concern of Brendan Cormier is not coming true: *We are still impressed by the 1748 Nolli map of Rome which traced a network of public spaces that included the interiors of churches, yet our urban plans today rarely try to project a public space network that links the interiors of our public institutions. And our architectural briefs rarely demand that this public element be included.* – "City Planning the Interior", *Volume no.33: Interiors*, 2012.

诺利地图，1748年
The Nolli map, 1748

展，同时冒着未经设计规划和自由使用的风险。这就好像数字世界的经验逻辑（提供服务以捕捉顾客，或至少吸引顾客的注意力）应用到了空间领域。新的苹果专卖店就是以这种方式运作的，试图通过提供对私有空间的自由访问来留住不同的客户，从而允许居民成为苹果世界的一部分。

从这个意义上说，所有呈现出来的项目都倾向于融合室内外空间的概念，以创建错综复杂、严格控制的空间；这种逻辑源于购物中心的演变：一个极其复杂的人造空间，在安全可控的环境中，到目前为止只有室内空间得以成功实现。

今天，物理界限的最小化，或室内空间的扩大，或室内外小气候日益得到综合控制，使得建筑物突破其局限性，室内外空间融为一个模糊的统一体。

这种发展趋势早已开始，如今已取得突出的成就。室内空间位于建筑内部，是人或物的本质；它是私有的、秘密的，没有边界或界限。（……）18世纪末和19世纪初的建筑，通常情况下，会封闭室内空间以获得舒适的个人空间。（……）灵活的装饰材料和完善的供暖和制冷技术使得这一切成为可能，但同时也产生了另一种可能：它连接室内空间和室外空间，因此几乎没有视觉边界。这种开敞式的室内空间，突然间开始威胁现代建筑的隐私性和保密性。——Rob Dettingmeijer，"展示隐私"，第33卷：室内设计，2012。

The projects presented below, indeed, show an exasperated research to push architecture through functions that require public, open uses, at the risk of *un-programmed and free for uses not yet conceived,* in the words of Farrell and McNamara. It is as if the experiential logic of the digital world (offering services to capture the customers, or at least, their attention) was transferred to space; the new Apple Store works in this way, trying to retain a varied clientele by offering free access to private spaces: this permits the inhabitants to be part of the Apple universe.

In this sense all the presented projects tend to mix the concepts of internal and external spaces, to create complex, but very controlled, universes; the logic derives from an evolution of the shopping mall: a space with conditions of extreme artificiality and complexity, but within a controlled and safe ambient that, until now, only the interiors succeeded in giving.

Today the limit becomes ephemeral and inside and outside can mingle in an indistinct continuum thanks to minimal physical separations, or to a scale enlargement of the interiors, or to the increasingly integrated control of the internal/external microclimate.

An evolution that starts from afar, but which today finds a definite fulfilment: *The interior is situated within. On the inside. It is the inner nature of a person or thing; private, secret, away from the border, or frontier. [...] Creating a physically sealed off interior as a zone of private comfort was one of the main efforts of late eighteenth and early nineteenth century building. [...] But the flexible insulating material and perfected heating and cooling techniques that made this possible also made the opposite possible: connecting inner and outer space almost without visual boundaries. Suddenly this act of open-*

大型商场步道，泰国
Mega Foodwalk, Thailand

山湖公园游乐场，美国
Mountain Lake Park Playground, USA

在这些项目中，我们可以看到室外空间移动到室内空间，或者室内空间融入了室外空间，室内空间的扩展改变了周围的景观。项目功能不断增加，目的是打造一个复杂而典型的城市有机体。

一个典型的例子是FOS建筑事务所设计的大型商场步道（28页），室内空间与自然景观相互交融，实现了人、自然和建筑的统一。建筑物的内部和外部相差无几，公共空间是一种建筑结构，一个有顶的开放式广场，人们在广场上可以俯瞰各个楼层及其楼梯、电梯和通道。

3XN建筑事务所和HKS建筑事务所设计的皇家剧场（80页）致力于在建筑物内部打造一个大型开放的多样化广场，创造一个顶级的会议场所和剧院，舞台上的表演则有助于将更多的人聚集在一起。

当代公共空间的新氛围，即室内空间和室外空间界线的模糊化，仅仅是表面上的，其特点是对空间、材料和比例的细节控制，尤其是对室内空间的控制。

奥尔胡斯海港浴场（162页）清楚地展示了这种方法：大型港口很难控制污染程度和浴场安全，丹麦建筑师设计的游泳池系统减少了部分室外空间，因界定行为而被转变为室内空间。因此，公共空间不需要重要的物理边界，只需要融入周围景观的一部分就具有了理想的娱乐性特征。

ing to the interior was threatening the recently winning architecture of privacy and secrecy. – Rob Dettingmeijer, “Privacy on display”, *Volume no.33: Interiors*, 2012.

In the projects presented here we can notice a flow of the exteriors towards the interiors, or, vice versa, an expansion of the interiors which modifies the surrounding landscape: it is constant to add functions, in order to create the typical densities of complex, urban, organisms.

This happens even more in the Mega Foodwalk (p.28) by FOS Foundry of Space, where we can also observe the incorporation of nature in the interior spaces, completing the triad among human body, nature and architecture. The interior and the exterior are totally indistinguishable and the public space is the structure around which architecture can grow: a covered, but open, square on which the different floors of the building overlook, full of stairs, elevators, passages.

Also the Royal Arena (p.80) by 3XN + HKS works on the concept of bringing a large, open and differently configurable square inside the building, to create one of the meeting places par excellence, the theater: here the performance that takes place on stage is able to act as an aggregator of a multitude of people.

This indistinction between interior and exterior spaces, which builds the new atmosphere of contemporary public spaces, can also be noticed in spaces that should be merely external: they are characterized by having a kind of details, a control of the spaces, materials and proportions, more typical of interior spaces.

The Aarhus Harbour Bath (p.162) by BIG clearly shows this approach: in a large environment, a harbour, in which it is difficult to control pollution levels and the safety of bathers, the swimming pool system of the Danish architect cuts

天理市车站广场CoFuFun，日本
Tenri Station Plaza CoFuFun, Japan

帝国商店，美国
Empire Stores, USA

杰克逊山湖公园游乐场（200页）不需要设置围栏，只通过巧妙使用人行道就划分出了有助于令人相遇的不同区域。连续性指的是环形，一种触发人群聚集的形式，该观点又让我们想到班纳姆反思：当代人喜欢临时使用公共空间，且让自己的居住空间围绕一个空间中心设置。

环形定义了天理市车站广场CoFuFun（188页）的空间形状，这并非偶然，该空间完全是一个室外空间，但是研究细节、颜色和材料后你会发现，这是一个典型的室内空间。它像是处在一个由白色家具界定大房间内，这些家具伴随人们运动的元素，采用离心力设计引导人们走上不同楼层，它们也是人们休息和社交的区域。有时，私有空间的内部会部分对外开放，与公共空间相连，形成通道，影响交叉空间的特征。

S9建筑设计事务所和STUDIO V建筑事务所设计的帝国商店（12页）就是如此。建筑朝向海景开放，使海岸和城市之间有了互动，鼓励人们的运动和好奇心。以为身处室内，实际却在室外所给人带来的惊喜展示了建筑内部的设计机制，创造了一种疏离的效果，形成了一个很难定义的模糊之地。公共主题就是在这种模糊的状态下被插入的，这使得城市环境更加复杂，也再次提高了人类的存在感。

在由克罗菲和麦克格雷戈·考克斯设计的梅特兰Riverlink（120页）项目中，建筑向社区开放，但是经过改造它本身变成了不同的建筑

out a portion of the exterior, transforming it, thanks to this act of defining a border, into an interior. The public space, then, acquires the ideal characteristics for recreational use, without the need for important physical boundaries, but only through the gesture of incorporating a part of the surrounding landscape.

The Mountain Lake Park Playground (p.200) by Bohlin Cywinski Jackson does not need fences, but through a wise use of pavements, it manages to delimit different areas that contribute to returning to an idea of encounter between people. The continuous referring to the circle, to a form that stimulates the aggregation of communities, brings back to Banham's reflections: the new contemporary nomads use public spaces temporarily, leaving their own residential spaces, around a spatial center.

It is not by chance that the circle defines the spatial shapes in the Tenri Station Plaza CoFuFun (p.188) by Nendo, a space that is totally external, but with a research on details, colors and materials, typical of an interior space. It seems to be in a large room, defined by a set of white furniture: they are elements that accompany people's movements, directing them with centrifugal forces, hosting them on several stairways, which become rest and socialization areas. Sometimes it happens that the interiors of private spaces partially open themselves, allowing the public space to enter, leaving a trace of its passage, influencing the characteristics of the crossed space.

This is what happens in the Empire Stores (p.12) by S9 Architecture + STUDIO V Architecture, which opens the architecture towards the water landscape, allowing a mix of flows between the shore and the city, encouraging people's movement and curiosity. The surprise of being inside, yet outside, of showing the internal organs of the architecture,

梅特兰Riverlink，澳大利亚
Maitland Riverlink, Australia

巴勒斯坦博物馆，巴勒斯坦
Palestinian Museum, Palestine

形式：它不再是碎片化的室内空间，而是一个大型避难所，一个保护人们、抵御恶劣天气的掩护场所。在这里，运动、通道也得到了强调：公共空间不仅是一个供人们沉思、久坐的地方，还是一个动态的场所，一种压力下的流畅空间，使人们从城市转移到自然，让水流回城市。通过公共空间改变空间作用，可以通过将内部功能向外界开放来实现。

这是一个典型的公共空间模式，博物馆的光环在城市环境中强烈散发，使运动场所与社会性概念直接相关。

赫尼根·彭建筑师事务所设计的巴勒斯坦博物馆（96页）是一座标志性建筑，在当地享有盛名。建筑师对它进行了大力改造，就像建筑为景观增色一样，这座建筑使自然和城市变形，这就如同石头掉进水里一样。它们像是由力产生的波，勾勒出一系列从中央建筑辐射到周边空间的同心公共空间。

赫尔辛基的阿莫斯·瑞克斯博物馆（60页）由JKMM建筑师事务所设计，其设计原理相似，但不是在周围形成一个力场，而是向外部扩散，使建筑物前方广场上晶体结构的有机体变得生动，而且设置了很多天窗。它们是公共空间的旅游伙伴，也是人们使用该空间的视觉和触觉焦点。

creates an alienating effect, an ambiguous place, difficult to define. In this ambiguity the public theme is inserted, which makes the urban surroundings more complex; it fosters, again, the presence of people.

In the Maitland Riverlink (p.120) by CHROFI + McGregor Coxall, the building opens to the community, but it is so strongly transformed that it becomes itself a different form of architecture: it is no longer a torn interior, but rather a large shelter, a cover that protects the meeting of people against adverse weather conditions. Here too the movement, the passage, are underlined: the public space is not only meditative, sedentary, but it can be dynamic, a fluid under pressure that moves people, taking them from the urban territory to the natural one, allowing the water to return to the city.

A way of appropriation of the territory, through the public space, can happen through the reverberation of internal functions towards the outside.

It is a typical modality of the public spaces par excellence, the museums, whose aura emanates strongly in an urban environment, or of the sport places, whose use is directly connected to a concept of sociality.

The Palestinian Museum (p.96) by Heneghan Peng Architects is an iconic building complex, well recognizable in its surrounding: it changes the surrounding forcefully, as irradiating the landscape by an architectural presence, which deforms nature and city, like a stone falling into water. The buildings are waves of force, which outline a series of concentric public spaces, from the central building to the peripheral spaces.

The Amos Rex Museum in Helsinki (p.60) by JKMM Architects works in a similar way, but instead of creating a force field around it, it seems to explode towards the exterior, giving life to a crystallized organism in the square in front of

©Rafael Gamo

绿湾Titletown公园，美国
Green Bay Titletown District, USA

©Vincent Fillon/Dominique Perrault Architecte / Adagp

巴黎隆尚赛马场，法国
Paris Longchamp Racecourse, France

这种形式也出现在ROSSETTI设计的绿湾Titletown公园（178页）中，综合体的主要建筑本身就是景观不可分割的一部分，长期的下降形成了一个人造地形，在自然、技巧、人性的混合中，模糊了室外和室内之间的界线。运动不仅是对身体的挑战，而且是一个复杂的行动世界，这使得建筑能够构建不同的场景，完美地为人类活动提供场地。

由多米尼克·佩罗建筑事务所设计的平台/建筑巴黎隆尚赛马场（144页）是体育景观的视觉焦点，平坦的地形成为设计主导，没有明显的立面。它是高高耸起的建筑物，为所有聚集在这里的人创造了一个坚实的避难所。看台似乎在向下滑动，向有人的地方移动，像是在欢迎他们，邀请他们上台，使用看台。用法雷尔和麦克纳马拉的话来说，随着时间的推移，建筑物本身就能找到与人分享和接触的方式。

因此，里希特建筑师事务所设计的佩科斯县安全休息区（134页）融合了景观和建筑，作为停靠点设置在场地上。它的存在改变了一个大型室内空间的景观。在这里，人们可以在旅途中享受片刻的休息和舒适，这与建筑的原材料直接相关。它是一个提供机会的空间，一种超越单一功能的关系，建筑因此定义了它的功能——它是公共空间中支持人际关系的一个要素，而这个公共空间这就是我们所居住的一个更大的室内空间——地球。

the building, full of skylights: they are travel companions of the public space, points of visual and tactile anchorage for the people who use that space.

This also happens in the Green Bay Titletown District (p.176) by ROSSETTI, where the main building of the complex is itself an integral part of the landscape: the long descent builds an artificial topography that dilutes the boundaries between what is external and what is internal, in a mix of nature, artifice, humanity. Sport not only is a moment of physical challenge, but becomes a universe of complex actions, which allows architecture to build different scenarios, that host perfectly human events.

The platform/building designed by Dominique Perrault Architecture, Paris Longchamp Racecourse (p.144), is the visual anchorage point of a sports landscape, dominated by a flat geography, without significant elevations: it is the building, then, to rise high in the territory, to create a solid haven for all the people who gather here. Sliding down, the stands seem to move toward the public, as to welcome them, to invite them to climb, to use architecture: in the words of Farrell and McNamara – *buildings themselves find ways of sharing and engaging with people over time.*

So, the Pecos County Safety Rest Area (p.134) by Richter Architects establishes an integration between landscape and architecture, placing itself as a point of docking in the territory: thanks to its presence, it transforms this landscape in a kind of large interior. Here people can find a moment of rest, of comfort during the journey, directly relating to the raw materials of architecture: a space that gives opportunities, a relationship that goes beyond the single functions; architecture thus defines its role, as an element that supports human relationships within a public space, which is then the great interior in which we live: the Earth.

梅特兰 Riverlink

Maitland Riverlink

CHROFI + McGregor Coxall

Mutual
MAITLAND PLAZA
JOHNSTONS

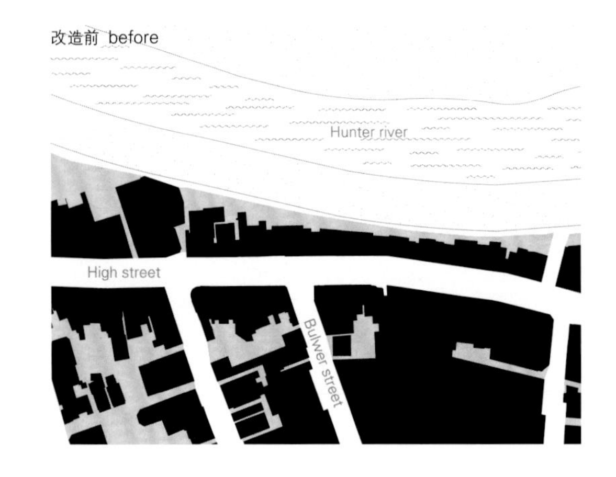

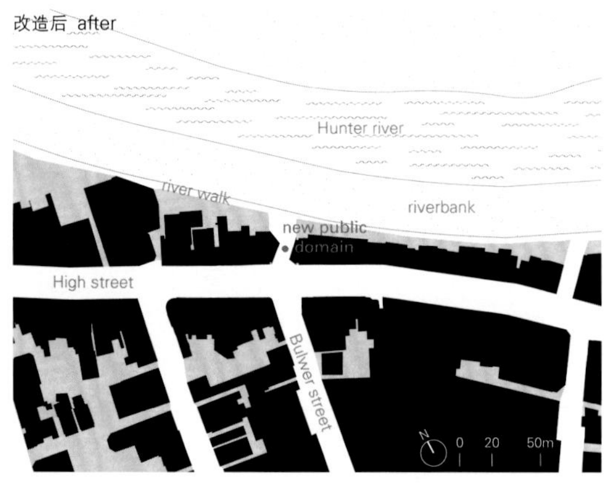

本案项目设计任务书的要求最初不包括公共空间，只包括咖啡馆和便利设施。最大化建筑物中央空间的举措进一步激活了公共空间和河滨区域。近年来，梅特兰的乡村城镇中心已经远离了河流，与主要的商业区和社区活动断绝了联系。洪水多发的灾难意味着当地人不再把河流看成是一种资产，而是一种对社区的威胁。建筑师发现了一个可以帮助人们重新构建这一概念的机会，并将建筑重新开发成为振兴城镇中心的枢纽。

这座建筑首次将梅特兰的两项重要资产结合在一起——一项是建筑风格丰富多样的商业街以及最近完成的公共领域升级改造，另一项是亨特河的环境舒适度。这个空间充当了社区的"公共起居室"，重新激活了城镇中未被使用的部分，并把当地人吸引回来。它是地区遗产的基本组成部分，同时也把游客带到了城镇。该结构可以转换为室外电影院或舞台，满足一系列社区活动。因此，梅特兰Riverlink项目将把小镇重新定位为一个充满活力的零售活动中心。

建筑师的意图是使Riverlink在重要建筑林立的街道上充当城市场景。它的高度与周围建筑的高度一致，因此虽然建筑形式大胆且为一个整体结构，但却让人感觉非常舒适。项目场地战略性地选在布尔沃街的末端，连通中心商务区南面的火车站。因此，从堤岸、邻近的洛恩大桥和贝尔莫尔大桥来看，它既是标志性的结构，又是"家具的一部分"。设计平衡了这些建筑方面的雄心和对人体规模的考量。"极具雕塑感的门户"构成了一个冬暖夏凉的公共空间，邀请人们使用。墙体、天花板和地板的精确角度使中央空间扭曲，巧妙地减缓了人们的行走速度。

建筑师选择砖块作为主要的建筑饰面材料，以补充原有街景甚至是梅特兰的建筑材料和尺寸。黏土砖的温暖和纹理在城市规模和室内的人体尺度上都发挥着作用，提供了一个耐久饰面。手工制作的特殊构件有助于整体砖墙弯曲成不太可能实现的角度，赋予材料意想不到的抽象特质。在需要轻质、透明或可开启材料的地方，建筑师利用轻质木条板组成的互补图案来提供这些品质，同时还保持了砖块图案的小尺度。触感丰富的材料让人们不禁想触摸建筑物。

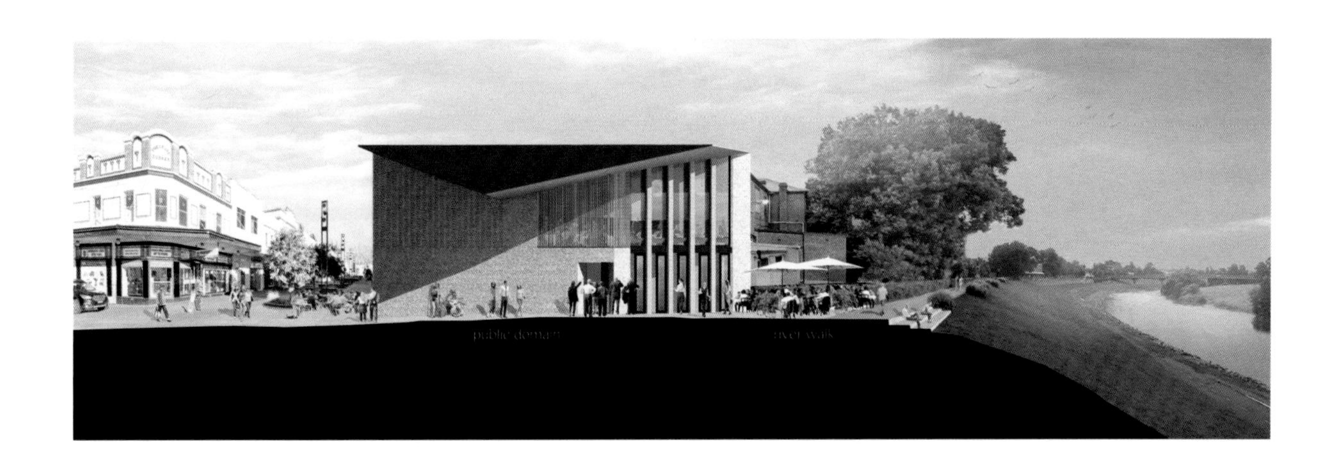

理想情况下，之前只是被当作商店背面的河滨现在邀请着人们在河流和街道之间自由活动。梅特兰市市长克里·洛雷塔·贝克说："我看到过一生都住在梅特兰的当地人，他们走到河边，眼睛闪闪发光，好像从来没有见过这条河。"

The brief for this project originally did not include a public space, but only a café and amenities block. The gesture to maximise the central space in the building further triggered an activation of public space and riverfront. In recent years, rural Maitland's town center had turned away from the river, disconnecting it from its main commercial and community activities. A series of devastating floods meant locals no longer see the river as an asset but as a threat to the community. The architects identified an opportunity to help reframe this notion and redeveloped the building into a pivot for the revitalisation of the center of the town.
The building unites Maitland's two key assets for the first time – its architecturally rich High Street with recently completed public domain upgrades and the environmental amenity of the Hunter river. The space acts as a kind of "public living room" for the community, reactivating an unused part of town and drawing locals back to the river, a fundamental part of regional heritage, whilst bringing tourists and visitors to the town. The structure can convert into an outdoor cinema or stage to accommodate a range

of community activities. Thereupon, Maitland Riverlink will reposition the town as a dynamic retail activity center. Riverlink intends to act as a civic set piece in the street full of great buildings. Its bold and monolithic form is yet read comfortably by maintaining the height of the surrounding forms. The location was chosen strategically at the end of Bulwer St, which links the train station to the south of the CBD. As such, it is delicately balanced to be both an iconic landmark when viewed from the levee bank, from neighboring Lorn and the Belmore Bridge, and also "a part of the furniture". The design balances these architectural ambitions with consideration for the human scale. The "sculptural gateway" frames a public space that invites occupation, a shady spot to sit in summer and a warm, protected place to enjoy the sun in winter. The precise angles of the walls, ceiling, and floor, twist and distort the central space, to subtly slow the movement through.

Brick was chosen as the primary building finish to complement the existing materials and dimensions of the streetscape and Maitland in general. The warmth and texture of clay bricks work at the urban scale and at the interior human scale to provide an enduring finish. Hand-made specials help the monolithic brick walls bend at unlikely angles giving the material an unexpected and abstract quality. Where there is a requirement for light, transparency or operability, a complementary pattern of light-weight timber batten panels is used to provide these qualities while retaining the fine-scale expression of the brick patterning. The tactile materials invite visitors to touch the building.

Ideally, the riverfront, previously seen as the backside of the shops, will invite people to move freely between the river and the street. Amazed by the project, "I have seen local people, who have lived in Maitland their whole lives, walk through to the link and their eyes light up as if they have never seen the river before", says Maitland Mayor, Cr Loretta Baker.

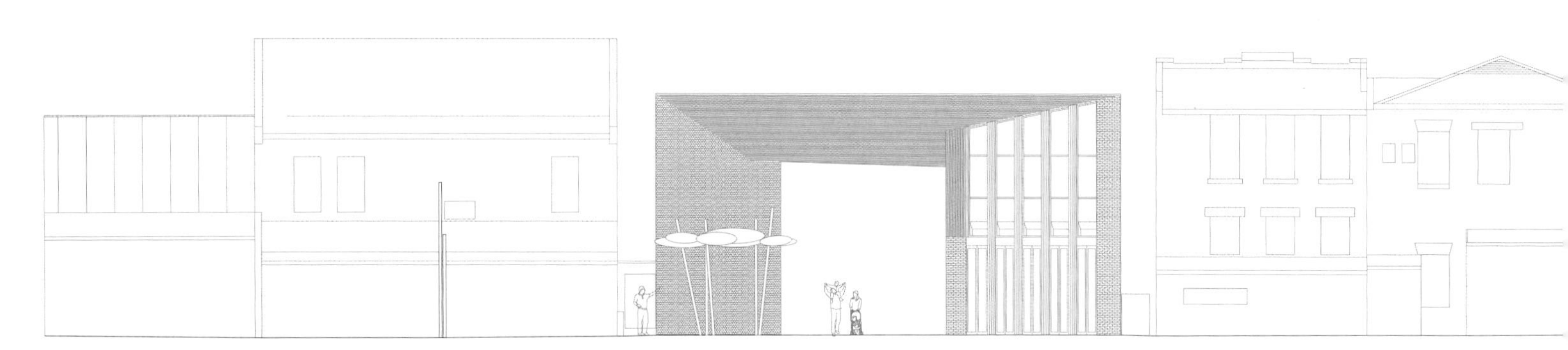

北立面 north elevation

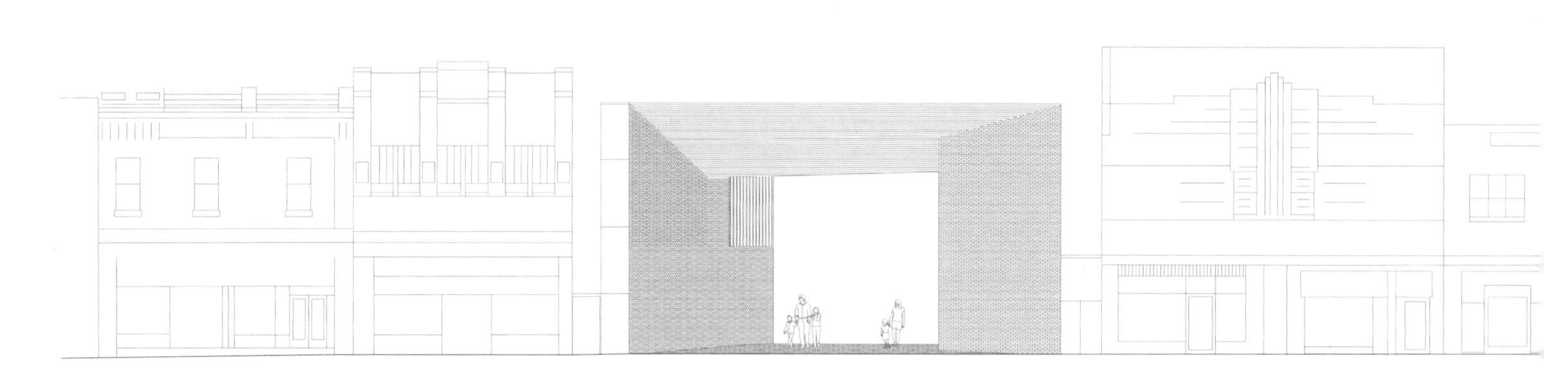

南立面 south elevation

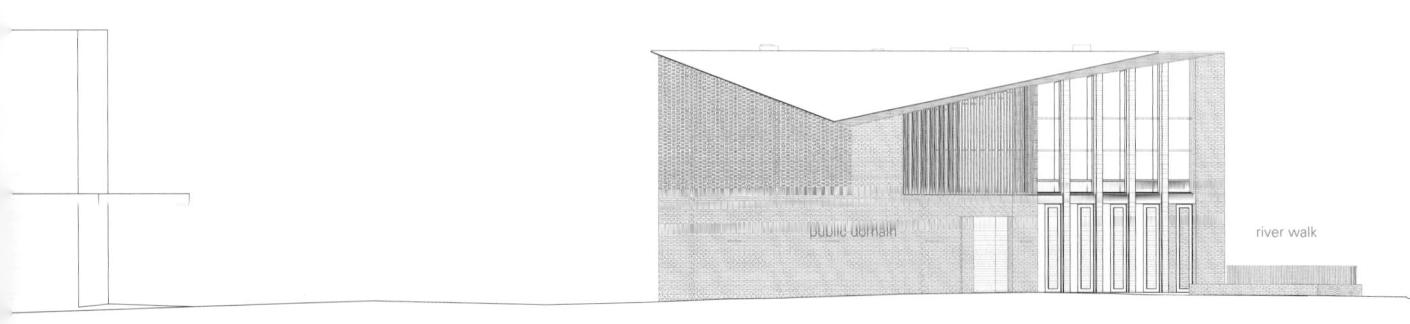

A–A' 剖面图 section A-A'

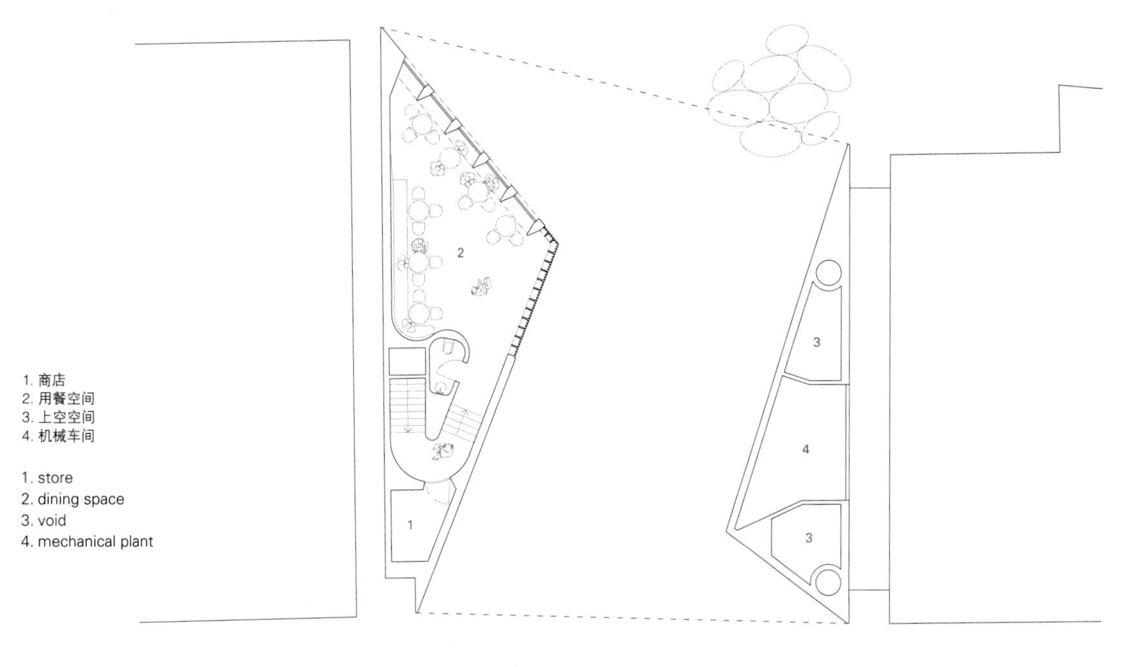
1. 商店
2. 用餐空间
3. 上空空间
4. 机械车间
1. store
2. dining space
3. void
4. mechanical plant

二层 first floor

1. 咖啡馆/酒吧
2. 厨房
3. 垃圾房
4. 男卫生间
5. 家庭房
6. 无障碍卫生间
7. 女卫生间
8. 视听室
9. 公共空间
10. 公共艺术设施
11. 河边散步道
1. café/wine bar
2. kitchen
3. bins
4. male WC
5. family room
6. accessible WC
7. female WC
8. AV room
9. public domain
10. public art
11. river walk
detail 1
A'
A
N
0
1
2m

一层 ground floor

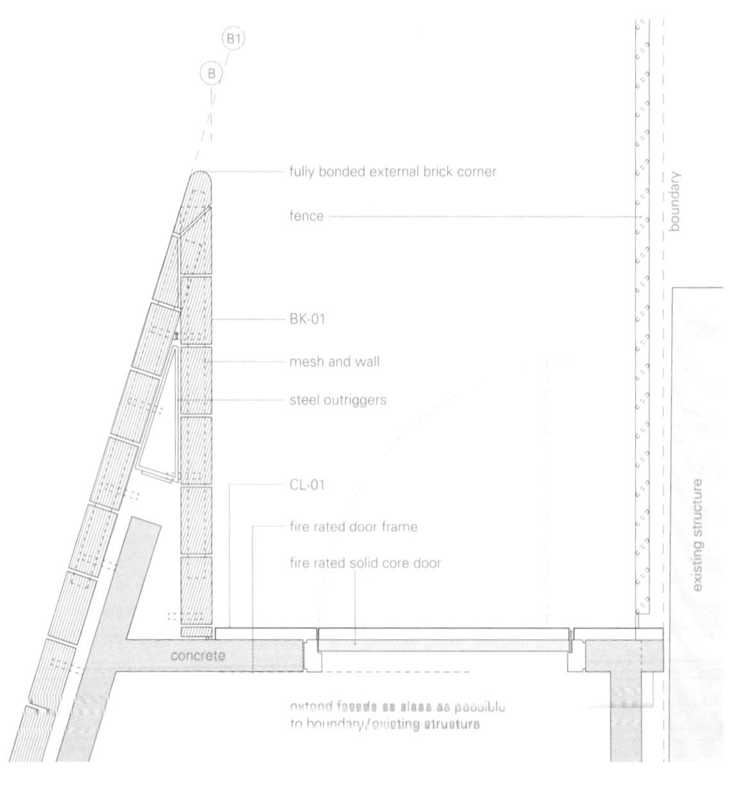
B1
B
fully bonded external brick corner
fence
boundary
BK-01
mesh and wall
steel outriggers
CL-01
fire rated door frame
fire rated solid core door
existing structure
concrete
extend facade as close as possible
to boundary/existing structure

详图1 detail 1

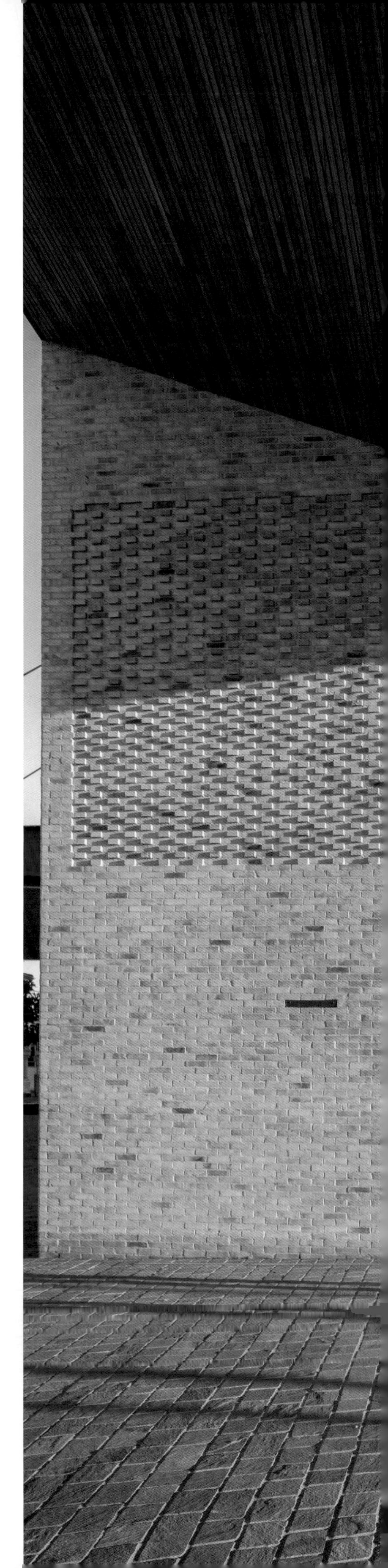

项目名称：Maitland Riverlink / 地点：396-400 High Street, Maitland, Australia / 事务所：Tai Ropiha, Joshua Zoeller, Susanne Pollmann-CHROFI; Adrian McGregor, Ann Deng, Julia Manrique, Maria Sabrià-McGregor Coxall / 结构工程师：SDA, Cardno
遗产：City Plan / 规划：JBA / 机械工程师：Northrop / 电气工程师：Northrop / 照明工程师：Northrop
水利工程师：Whipps-Wood / BCA：Steve Watson Partners / 通道：BCA Access Solutions / 剧院、视听室：Marshall Day Entertech
厨房：Frost / 工料测量工程师：RLB / 公共艺术设施：Braddon Snape / 防火工程师：MCD Fire Engineering / 客户：Maitland City Council / 用地面积：439m² / 建筑面积：147m² / 竣工时间：2018 / 摄影师：©Brett Boardman (courtesy of the architect)-p.126, p.128~129[right], p.130~131[left]; ©Simon Wood (courtesy of the architect)-p.123, p.124, p.125, p.132~133; ©Edge Commercial Photography (courtesy of the architect)-p.120~121, p.122, p.131; ©Matthew Abbott (courtesy of the architect)-p.128

佩科斯县安全休息区

Pecos County Safety Rest Area

Richter Architects

佩科斯县安全休息区位于美国10号州际公路旁，在那里可以看到戴维斯山国家公园。这个安全区可为人们提供必要的旅行便利设施，它的设计重新诠释了当地景观。天然的草甸、平顶山和山脉沿着广阔的远处景观分层堆积。高速公路沿线会时不时显露出原生石灰岩地层。这座建筑的核心灵感来自无尽的山脉、广阔的天空和展示这些美的方式——这里的美虽然可以在高速行驶的汽车上欣赏，但是却只有在地面上行走时，才能欣赏到全部的美。无论是从尘土和岩石中隐约可见的一簇簇花朵，还是层峦叠嶂勾勒出的远方地平线，都在讲述着这片土地过去、现在和未来的故事。

该项目的设计旨在满足功能需求，给予旅行者更好的体验感。该项目的基本目标是提供公共卫生间、自动售货机、文化和历史展品、野餐区、游乐区、自然小径和供舟车劳顿的司机休息的停车场，这些司机已经沿着10号州际公路行驶了很久，需要休息，10号州际公路位于美

国南部，长达4000km，主要为东西走向。该项目包括东面和西面两套设施，以便更好地接待旅客。客户——德克萨斯交通部的目标是通过为驾驶员及其家人提供一个温馨、干净、引人入胜的休息和缓解疲劳的地方来提高公路的安全系数。设计中使用的材料包括天然采石场开采的当地石灰石、玻璃、耐候钢和木材。建筑的主要结构是带钢框架的承重砌体。旅行者乘坐汽车、货车、房车、大型半挂车或公共汽车而来。游客来源近至德克萨斯州及邻近州，远至加利福尼亚州、佛罗里达州、加拿大和墨西哥。

"建筑自身的对比、纹理和体量都会影响人的情绪。它甚至不用借助美丽的山脉来吸引你，其自身的美丽足以令你驻足。当你走近时，钢铁的颜色和斑驳的痕迹会激发你的想象……正是这些元素可以让人们在回到旅途之前清醒过来。"——德克萨斯交通部的安全休息区项目经理Andy Keith这样说道。

项目简介

德克萨斯交通部佩科斯县安全休息区设在美国10号州际公路的两侧，每栋建筑700m²（一栋在东面，另一栋在西面），由石头堆积而成，建筑稍作旋转，可以让人欣赏到远处的山脉景观，关注东西两侧的景色，东和西暗喻着起点和终点。建筑的几何形状有利于使用被动式太阳能。当地的石头错落有致地排列堆砌起来，象征着随着地形绵延起伏的高速公路上所裸露出的地质状况。走廊的屋顶随意折叠以与地平线呼应。木质地板让人想到德克萨斯州早期边远地区的建筑结构。落在建筑上的雨水会流入干枯的小溪。小溪与由耐候钢板搭建的野餐棚相连，提醒着人们水是如何滋养和雕刻着这片沙漠大地的。本土的沙漠植物生长旺盛，在项目施工过程中，它们通过栅栏围挡得到了保护。

With Davis Mountains in view, Pecos County Safety Rest Area sits along U.S. Interstate 10 providing essential travel amenities via substantive interpretation of the region. Natural grasses, mesas, mountains are layered along an expansive and remote landscape. Native limestone strata is periodically revealed along the highway. The core inspiration for the design of this project was the land itself, the big open sky, and many ways their beauty is revealed – beauty that can be sensed at highway speed but only fully grasped on foot. From the intimate scale of tiny flowers peeking from dust and rock to the distant horizon scribed by layered mountains – the land is the story of past, present and future here.

The design sought to elevate the travelers' experience while meeting a pragmatic and modest program. The basic purpose was to provide public restrooms, vending, cultural and historical exhibits, picnic areas, play areas, nature trails, and parking for tired drivers who travel along Interstate 10, a 4,000 km long major east-west highway in southern United States. The project includes an eastbound and a westbound facility to better accommodate travelers. The goal of the client – Texas Department of Transportation (TxDOT) – is to enhance highway safety by providing drivers and their families with a welcoming, clean, and engaging place for rest and relief from road fatigue. Materials in the design include natural quarried regional limestone, glass, weathering steel, and wood. Primary structure is load-bearing masonry with steel framing. Travelers come by cars, vans, RVs, large

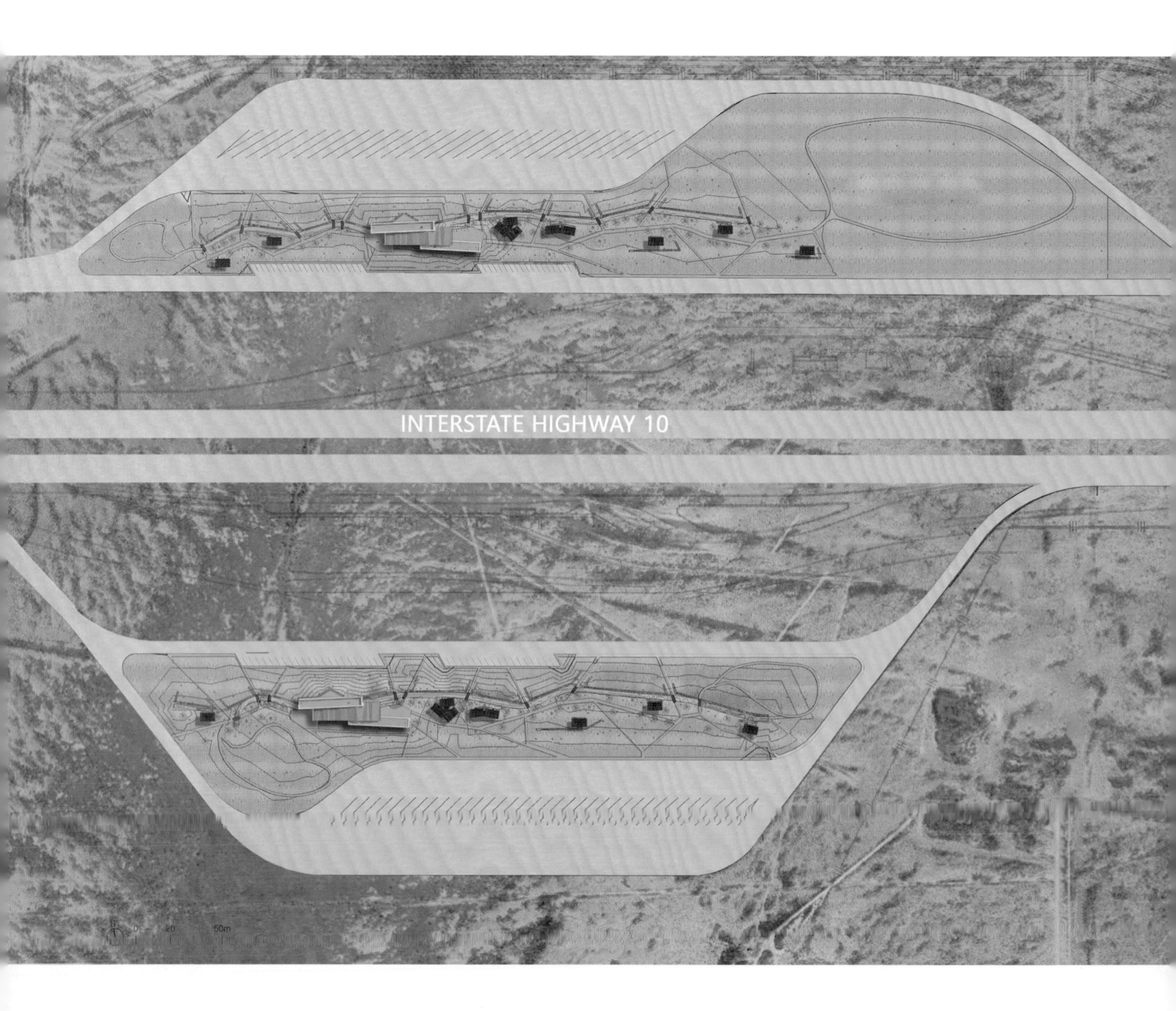

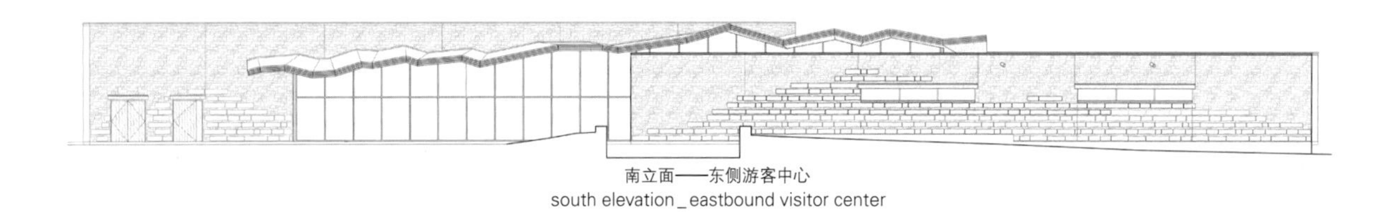

南立面——东侧游客中心
south elevation_eastbound visitor center

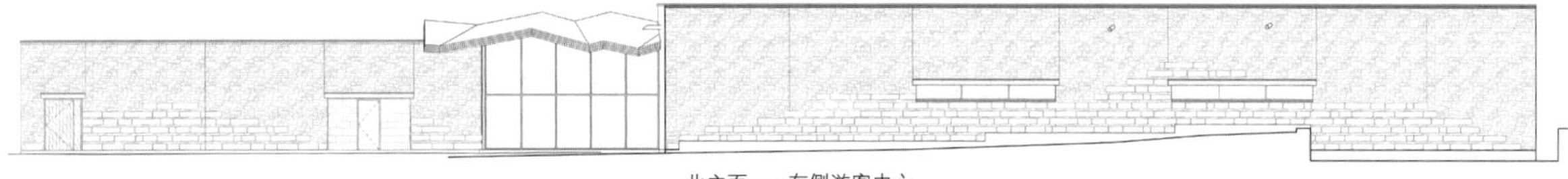

北立面——东侧游客中心
north elevation_eastbound visitor center

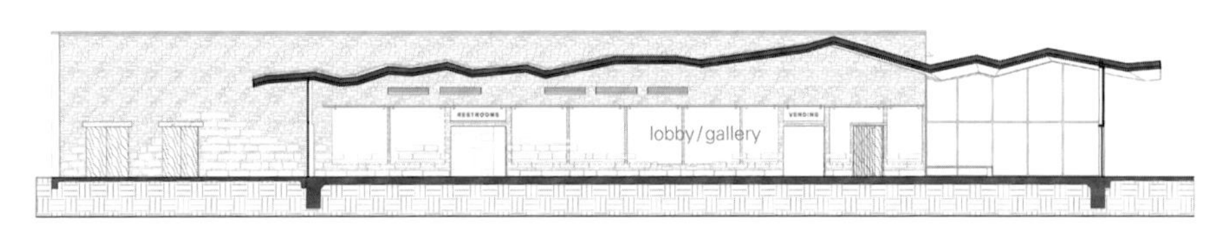

A-A' 剖面图 section A-A'

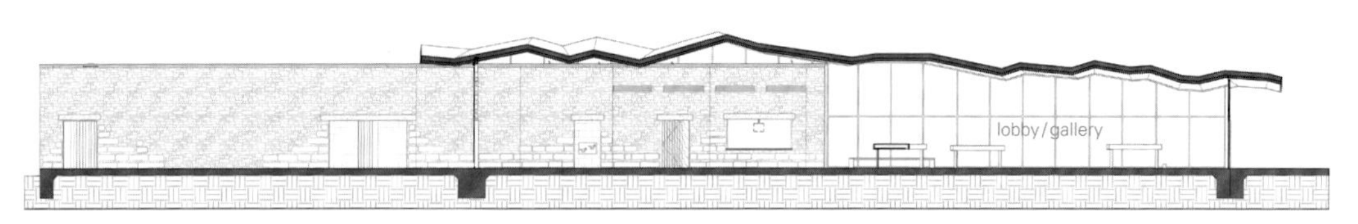

B-B' 剖面图 section B-B'

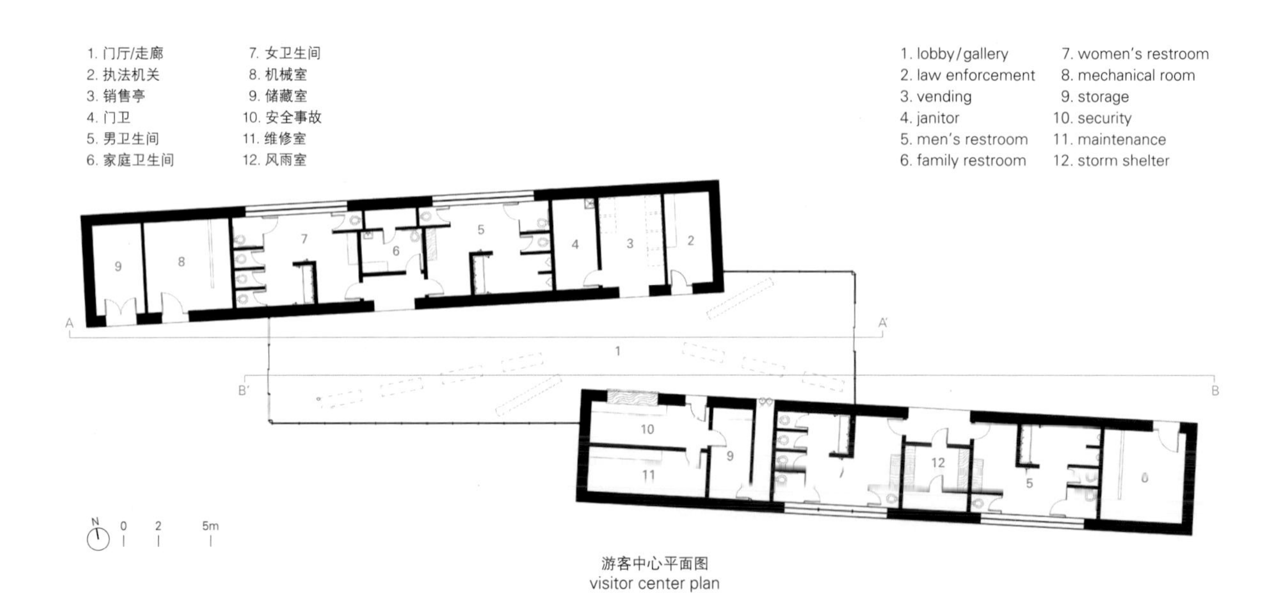

游客中心平面图
visitor center plan

semi-trailer trucks, and buses. Visitors come from Texas and neighboring states and as far away as California, Florida, Canada, and Mexico.

"The contrast, texture and massing have a way of influencing mood. It doesn't even need the beautiful mountain settings to stop you in your tracks. When you walk up close, the colors and imperfections in the steel stir the imagination... That's exactly what get people to wake up before they get back on the road." – Andy Keith, TxDOT Safety Rest Area Program Manager

Project Synopsis

The TxDOT Pecos County Safety Rest Area consists of a building and site amenities along each direction of U.S. Interstate 10 freeway. Each building at 700m² (one eastbound and one westbound) is composed of two stone masses shifted to frame borrowed views of the mountain horizon and to focus through-views to the east and west – metaphorically the origin and destination. The geometry further provides beneficial passive solar orientation. Native stone is organically coursed to suggest the geology revealed in highway land cuts as topography rolls. The roof of gallery spaces randomly folds to echo the horizon. Wood plank floors give audible recall to early Texas frontier structures. Dry creeks carry rain water away from the buildings while connecting the weathering steel plate picnic shades with reminders of how water nourishes and carves the desert land. Native desert plants thrive and they were fenced and protected during construction.

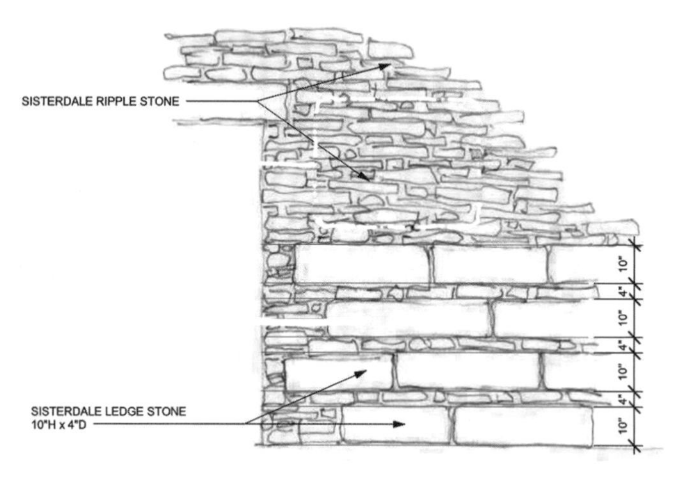

标准建筑石材饰面图案立面
typical building stone veneer pattern elevation

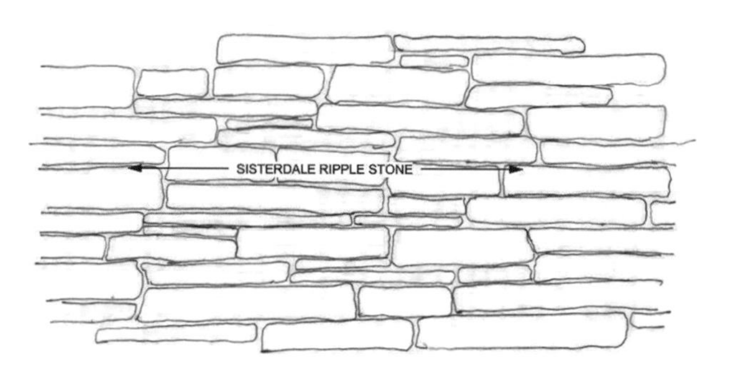

标准波纹饰面石材图案
typical ripple veneer stone pattern

项目名称：Pecos County Safety Rest Area
地点：Interstate 10, West Pecos County, Texas 79735, USA
建筑师：David Richter, Elizabeth Chu Richter - Richter Architects
项目团队：Sam Morris, Stephen Cox, Bob Mitchell,
Aaron Geiser, Lonnie Gatlin, Albert Delgado Jr, Toan Q. Huynh
用地面积：49,011m²-eastbound; 69,919m²-westbound
建筑面积：706m²-eastbound; 706m²-westbound
竣工时间：2018.2
摄影师：
©Craig Blackmon (courtesy of the architect) - p.137, p.138~139, p.141, p.142~143
©David Richter (courtesy of the architect) - p.134~135
©Elizabeth Chu Richter (courtesy of the architect) - p.140

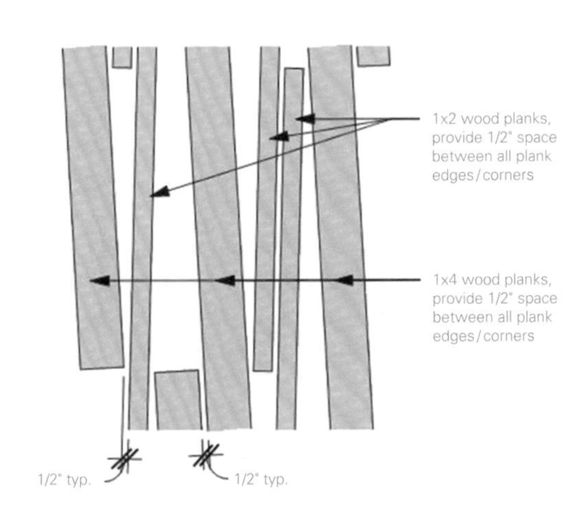

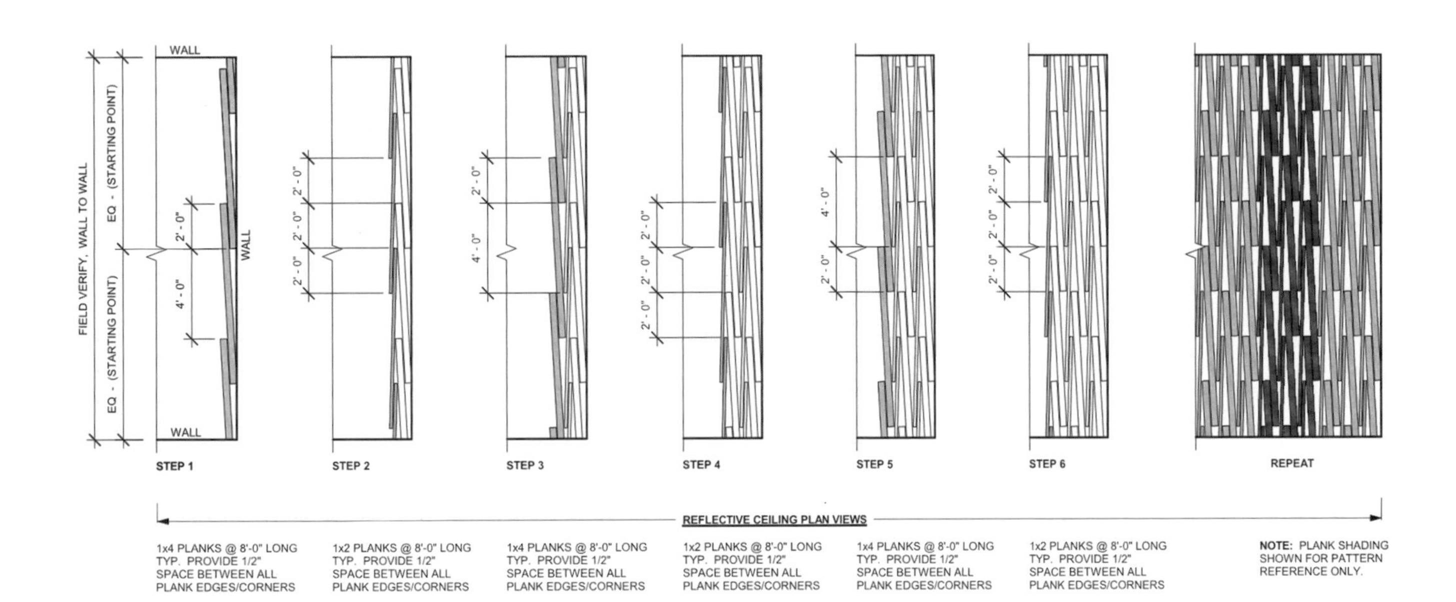

巴黎隆尚赛马场

Paris Longchamp Racecourse

Dominique Perrault Architecture

经过两年工程浩大的建设，隆尚赛马场终于重新开放。

这座历史悠久的赛马场坐落在布洛涅森林的中心，由法国建筑师多米尼克·佩罗设计，决定打破传统赛马场的贯有规范，向公众开放。

佩罗的设计是采用景观项目的形式重建整个区域，突出建筑环境及赛马场的历史遗迹。翻新后的跑道及设施使隆尚赛马场仍是世界上最受推崇的赛马场地之一。

每年，隆尚赛马场都会举办著名的凯旋门大赛，这是欧洲参赛人数最多的赛马比赛之一。设计团队面对的最大挑战是要设计出既能够容纳近60 000观众，又要在普通比赛日容纳少数观众的活动场地。因此，建筑师提出拆除20世纪60年代以来修建的所有看台，并用一个更紧凑且更具功能性的大看台取而代之。该建筑共四层，约10 000个座位，包括顶楼的餐厅和屋顶露台、五间接待室、五间酒吧、一间小酒馆以及媒体和酒店设施。下面两层向公众开放，第三层供专业人士和业主使用，第四层和第五层为VIP区域。

该建筑以透明且没有前后的"架子"为设计概念，这样观众能够来回走动看到马厩和赛马场的景色。观众不需要直接接触便能在视觉上接近赛马与运动员。建筑设计简洁而优雅，开放式的楼层布局允许人们在不同的空间自由走动。

看台建筑采用动态的设计，仿佛一匹奔驰的骏马，而且将观众的注意力吸引到赛道上。看台层层偏移，最顶层看台向外悬挑约20m。轻微的悬挑使得叠加的看台在终点线的方向上形成互动。引导移动方向的"流畅建筑"以看台的悬挑为标志，并将景观引入其中。宽大的窗户引入了大量的自然光，给人以空旷和开放的感觉。

基础设施由混凝土和金属构成，看台用木头覆盖。呈水平状态的看台和金色铝立面在不同季节折射出不同的光线，使建筑与环境融为一体。丝网印刷的玻璃栏板与隆尚赛马场之前的看台相呼应，让人想起前赛马场的花卉植物。

巴黎隆尚赛马场恢复了昔日游园会人们在树林中散步、在巴格特尔花园中吟诗作赋的魅力，同时为各种不同的观众提供了功能齐全的设施。赛马场将会逐渐成为更接近自然的场所：一个更生态化的场所，被动式和主动式太阳能系统鼓励使用可再生能源和建筑物的能源独立；一个更加灵活的场所，通过创造多样的场所，让隆尚赛马场适应不同条件，容纳更多的人；一个更舒适的场所，让场地的所有用户感到舒适，包括赛马、骑师、男女马术员、专业人士和全体公众。新看台比之前的更低更短，与自然完美融合，该项目将是巴黎市政府制定的气候计划同类项目环境设计的学习目标。

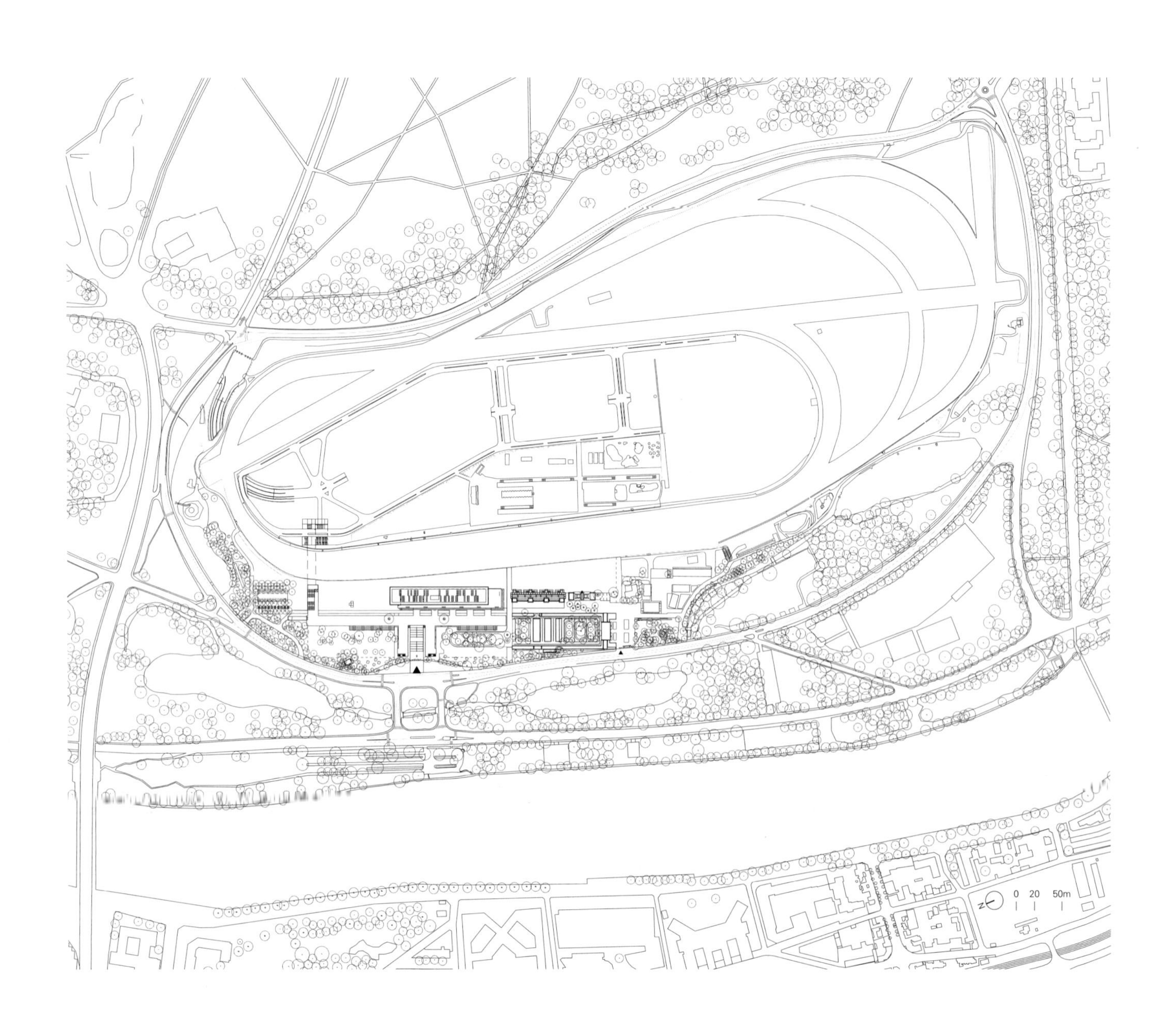

After two years of gargantuan building work, Paris Longchamp racecourse reopened its gates.
Designed by French architect Dominique Perrault, this historic horseracing venue nestled in the heart of the Bois de Boulogne has decided to break with the classic codes of traditional racecourses and open up its ground to the general public.
Perrault's plan took the form of a landscape project that transforms the entire area, highlighting the legacy of its built environment as well as the racecourse's historical heritage. The renewal of the course and its facilities ensures that Longchamp maintains its status as one of the world's most revered horse racing destinations.
Each year, Longchamp hosts the prestigious Prix de l'Arc de Triomphe – one of Europe's most attended horse races. The main challenge for the design team was to develop a scheme that could accommodate up to 60,000 spectators, while also welcoming smaller crowds on ordinary race days. Consequently, the project proposed to remove all the stands dating from the 1960s, replacing them with a single grandstand that is more compact and functional. With a capacity of about 10,000 seats, the building has four levels, including a restaurant with a rooftop terrace on the top floor, five reception rooms, five bars, a brasserie, facilities for the press and hospitality areas. The first two levels are open to the public, the second floor is reserved for the professionals and owners, the third and fourth floor for the VIP.

拆除
1. 村庄建筑
2. 露台和收银台
3. 市政委员会看台

重建
4. 成绩计算区
5. 总统将台
6. 骑师俱乐部看台
7. 亮马圈

保留
8. 亭子看台

翻新
9. 赛马赌金计算区
10. 养马场
11. 行政管理区

重建-新建
12. 养马场（扩建）
13. 成绩计算亭
14. 荣誉亭
15. 骑师俱乐部看台
16. Suresnes亭
17. P2图腾终点线
18. 行李传送区
19. 轨道楼梯

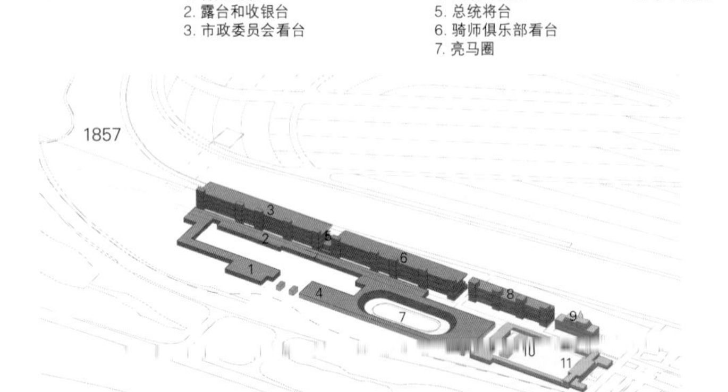

demolished
1. village building
2. terraces and cash desks
3. grandstand of the municipal council

rebuilt
4. scales
5. presidential rostrum
6. jockey club grandstand
7. parade ring

preserved
8. pavilion grandstand

refurbished
9. totalizer
10. stables
11. administration

rebuilt-built
12. stables(extension)
13. scales pavilion
14. honorary pavilion
15. jockey club grandstand
16. pavilion of Suresnes
17. finish line totem P2
18. carousel pavilion
19. tracks stairs

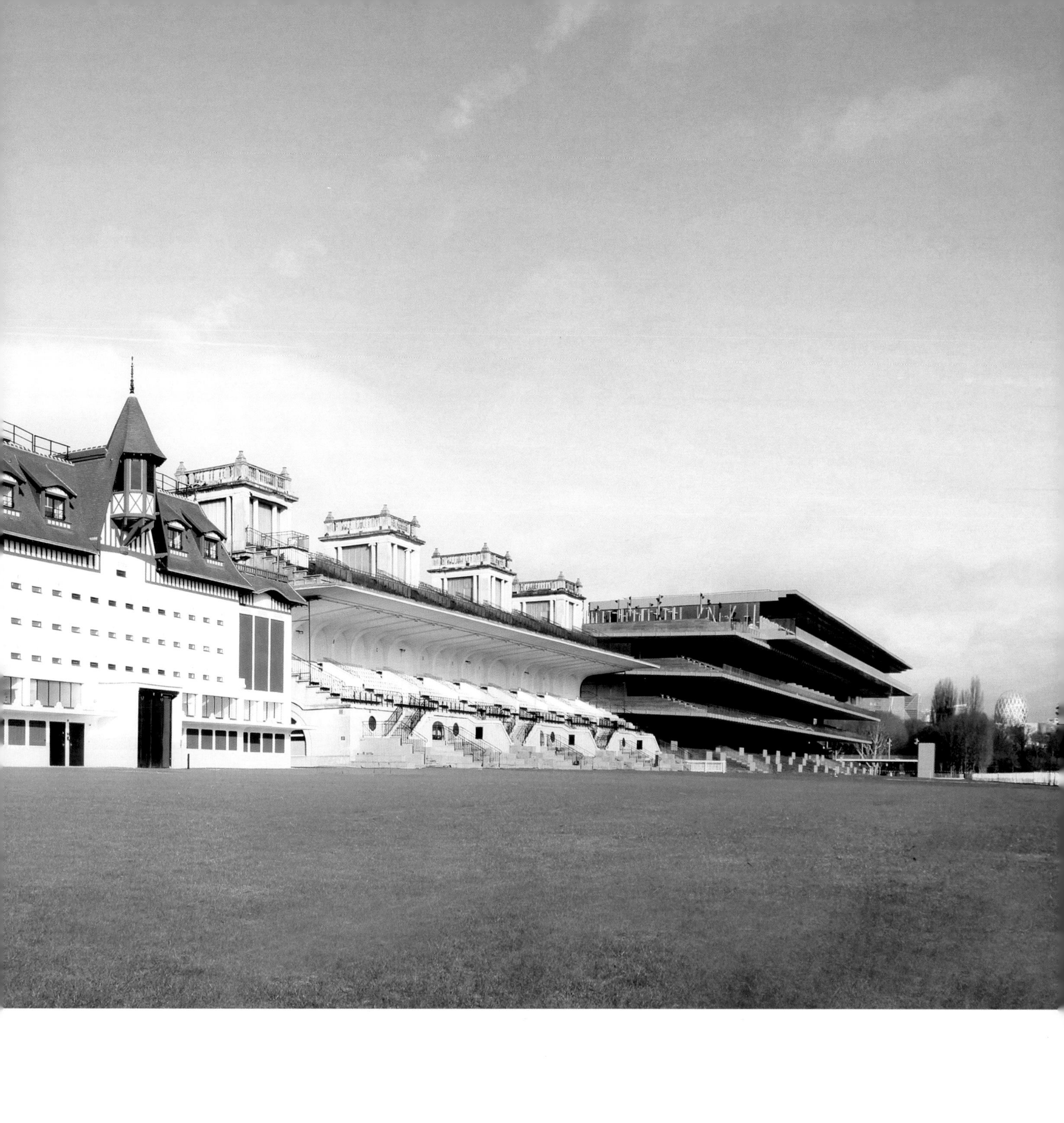

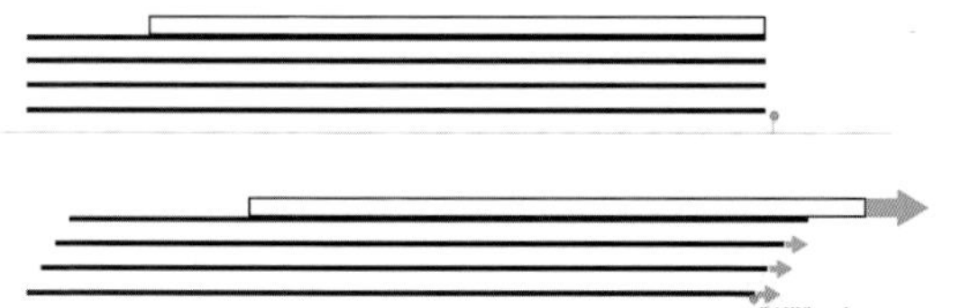

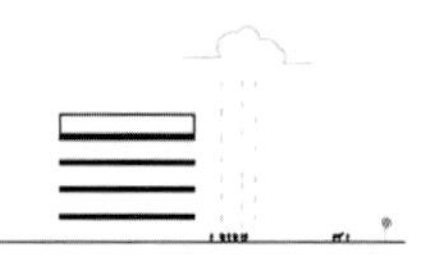

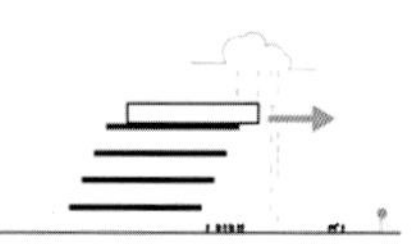

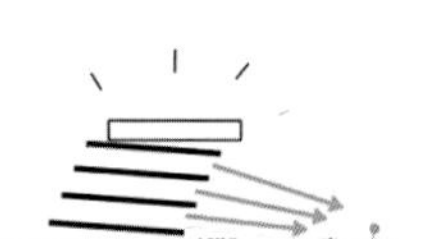

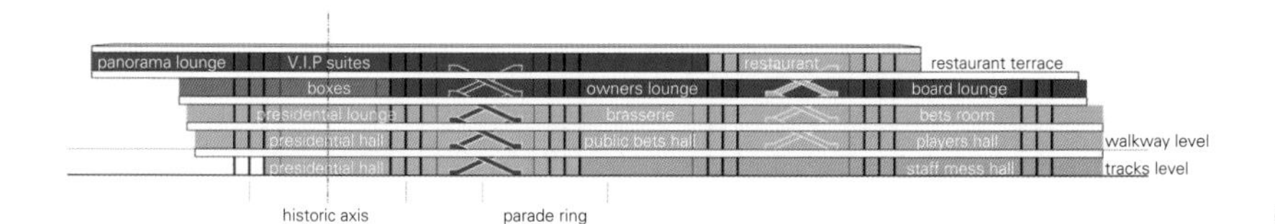

press
comissioners

presidential
spaces

V.I.P
owners

logistics
staff

public

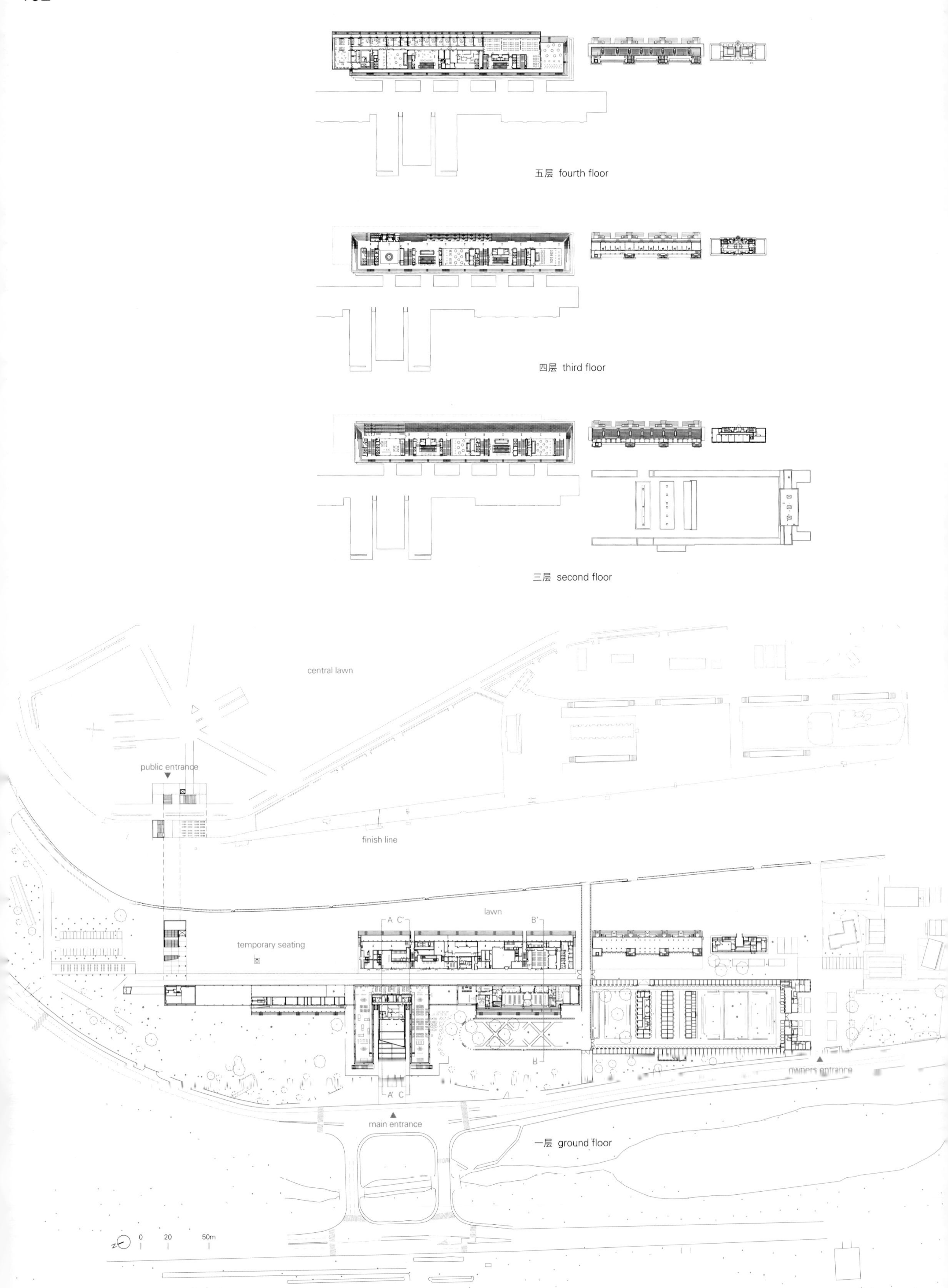

五层 fourth floor

四层 third floor

三层 second floor

一层 ground floor

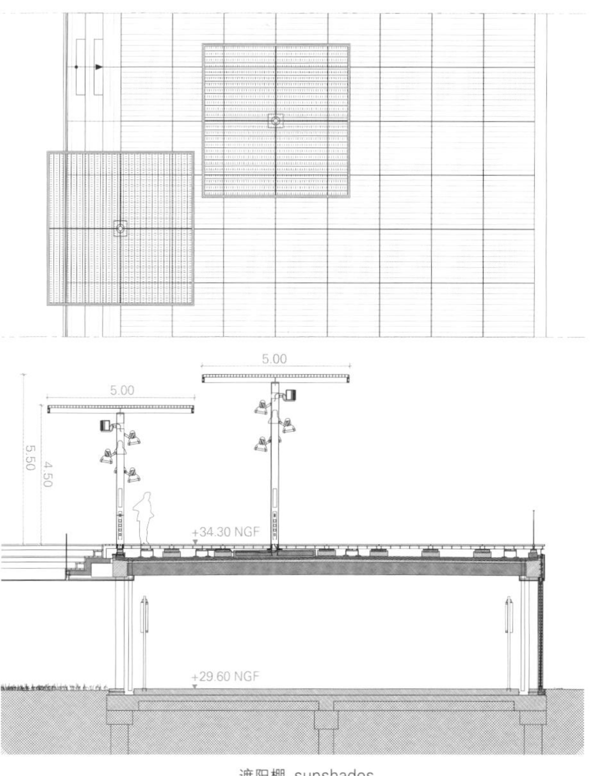

遮阳棚 sunshades

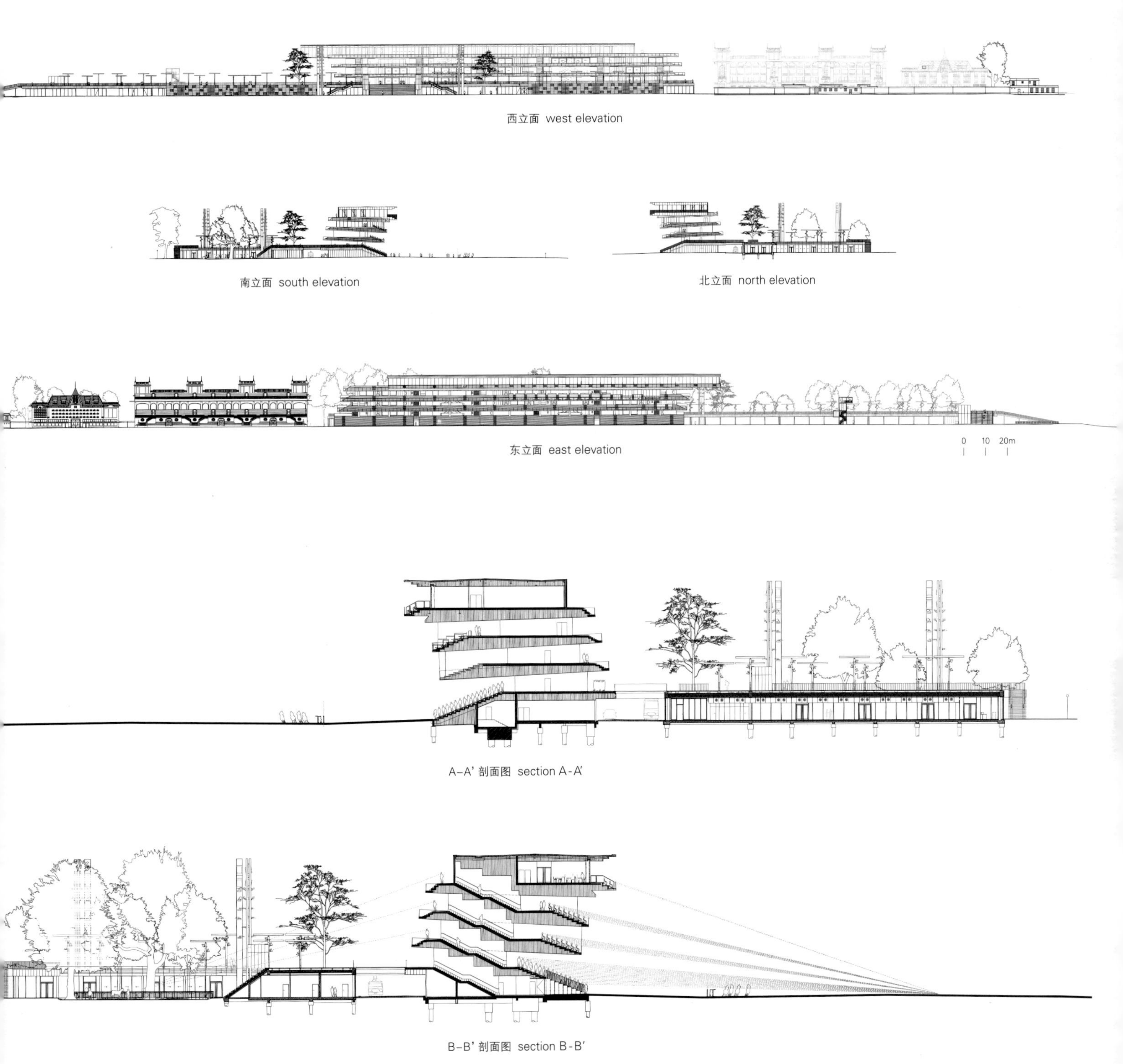

西立面 west elevation

南立面 south elevation

北立面 north elevation

东立面 east elevation

A-A' 剖面图 section A-A'

B-B' 剖面图 section B-B'

C-C' 剖面图 section C-C'

项目名称：Paris Longchamp Racecourse
地点：2, Route des Tribunes, 75116, Paris, France
事务所：Dominique Perrault Architecte
工程部：economist - RPO; museography - Renaud Pierard; structures - Tractabel Engineering; fluids - Oteis; facades - Terrel; acoustic, lighting - JP Lamoureux; landscape architect - Ter
用地面积：630,000m²
建筑面积：15,000m² SDO / 60,000m² SHOB
新骑师俱乐部看台面积：34,000m²
新骑师俱乐部看台体积：160m x 35m x 23m
草坪面积：55,000m² (35,000m² added)
光伏板面积：600m²
LEED认证：HQE - regulatory obligation RT2012 for the new buildings
供暖：100% of geothermal energy
研究开始时间：2012.9
施工开始时间：2015.10
竣工时间：2018.1 / 开放时间：2018.4.29
摄影师：
©Adagp (courtesy of the architect) - p.144~145, p.147, p.148~149, p.154, p.156~157, p.160, p.161 bottom
©Vincent Fillon_Adagp (courtesy of the architect) - p.150~151, p.153, p.158, p.159, p.161 top

The architectural concept was the transparent "shelf" without front or back, enabling spectators to go back and forth from a view over the stables to a view over the racecourse. The public is always in visual contact and proximity to the horses and professionals, without ever coming into direct contact. Designed simple and elegant, it comprises open floors in which the circulations allow free movements in various spaces.

In a dynamic movement, the tribune reminds of a galloping horse and its volumetry draws the attention down to the track. The levels are slightly offset from one another until the top level overhangs about twenty meters. A slight overhang orients the interplay of superimposed stands toward the finish line. The "fluid architecture" that guides the

movement, symbolized by the overhang of the grandstand allows the landscape to pass through it. The sensation of space and the feeling of openness are provided by a large amount of natural light through the generously dimensioned windows.

The infrastructure is made of concrete and metal, the stands are covered with wood. The horizontality of the tribune and the golden aluminum facades integrate the volume with the environment by playing with the light according to the seasons. Echoing the former tribunes of Longchamp, the screen-printed glazed balustrades evoke the flower planters of the former racecourse.

The ParisLongchamp racecourse revives the charm of garden parties of the past, of promenades in the wood and the poetry of the Bagatelle gardens, while offering an efficient facility for all types of public. It will gradually become a more natural place that is: more ecological, when both passive and active systems encourage the use of renewable energy and the energy independence of its buildings; more flexible, by creating a wide variety of places that welcome greater numbers under conditions adapted to the diversity of racecourses; and finally, more pleasant, for the comfort of all users of the site, the horses, jockeys, horsemen and women, professionals and the public at large. Forming a coherent whole with nature, the new tribune, lower and shorter than the previous one, will be an environmental model of its kind and thus meets the objectives of the climate plan developed by the city of Paris.

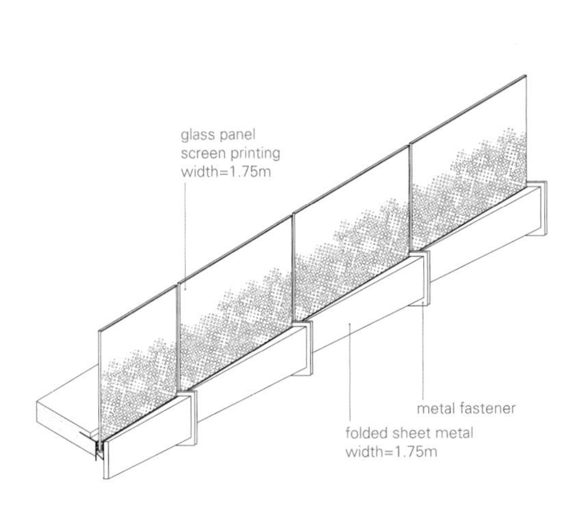
glass panel
screen printing
width=1.75m
metal fastener
folded sheet metal
width=1.75m

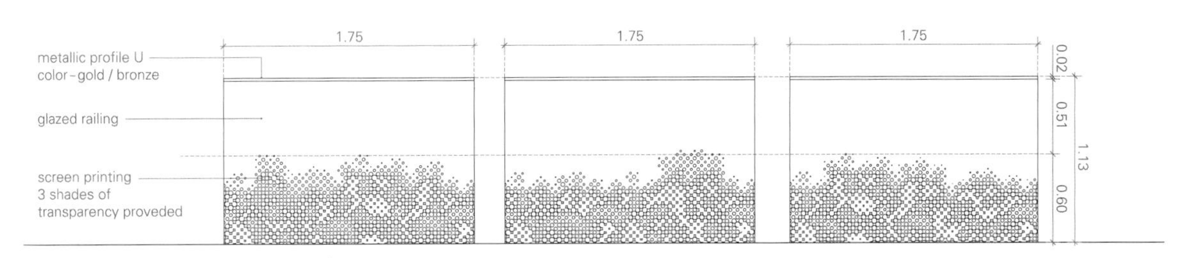
1.75
1.75
1.75
metallic profile U
color-gold / bronze
glazed railing
screen printing
3 shades of
transparency proveded
0.02
0.51
1.13
0.60

奥尔胡斯海港浴场

Aarhus Harbor Bath

BIG

趁阳光正好，或在海里畅游一番，或只是慵懒地躺下享受阳光。人们迎海风而行，伴夕阳而食。奥尔胡斯海港浴场位于新海滨区域，被称为"奥尔胡斯Ø"，宛若一片充满活力的绿洲。BIG建筑事务所与城市生活专家Jan Gehl共同设计并开发了这个多功能的浴场。奥尔胡斯海港浴场的两端就像是一对从港口突出的交错长廊，一直从Nikoline Kochs Plads小镇广场延伸到海滨区域，将城市与海湾连接起来。奥尔胡斯海港浴场与它所毗邻的海滩浴场全年对外开放，为大众提供新颖有趣的水上娱乐活动。BIG设计的第一个海港浴场于2002年在哥本哈根建成，哥本哈根也因此被称为世界上最宜居的城市之一。该浴场紧邻公园以及海港船只坡道，它得到国际奥林匹克委员会公布的2007年度最佳公共娱乐设施荣誉提名。

奥尔胡斯海港浴场规模更大、设计更宏伟，延续了BIG的一贯风格，即以最少的建筑物创造最大限度的生活空间。游泳者可以选择圆形跳水池、儿童泳池、50m的长形泳池或任意一个桑拿浴室。桑拿浴室上方是公共木板人行道，适合那些喜欢保持身体干爽的人，他们可以在此俯瞰泳池和大海景观。

人们在海港浴场不仅可以游泳，还可以参加各种各样的活动。浴场前方有一系列的独立餐厅、沙滩排球场、儿童剧院、形式各异的海滩小屋以及其他以创造生活为主题的公共项目。未来几年，包括名为"奥尔胡斯"的大型会议中心在内的私人建筑也将兴建。因此，与"奥尔胡斯Ø"同高的私人住宅建筑以及位于中央的庭院广场成为公共空间的一部分，它满足了人们对公共空间的需求，与近日完工的公共图书馆共同成为一道迷人的城市景观。以设计公共空间为第一步的总体规划将公共项目与私人住宅设计巧妙结合起来，从而创建了一个新的动态城市空间，在这里，公共空间与私人领域汇聚在一起。

奥尔胡斯海港浴场与其同类型浴场相比，拥有最为丰富的海水结构，最多可容纳650人。该浴场每周末开放，但其中的长廊甲板是每天24小时全年开放。奥尔胡斯作为丹麦人口第二多的城市，其海港浴场将为岛上的居民和游客提供更具吸引力和冒险性的滨水体验，它通过将水滨区域作为公共空间使用，为历史上保留用于工业目的的区域注入了新的生命。

Go for a swim, lie down under the sun, walk with the sea breeze, eat and drink with the sunset. The new Harbor Bath emerged as a vibrant oasis in the new waterfront neighborhood named Aarhus Ø (Aarhus Docklands), BIG's current mixed-use development plan with fellow urban life expert Jan Gehl. It is a pair of overlapping promenades jut out from the harbor, stretching from the Nikoline Kochs Plads town square to the tip of the waterfront, binding the city with the bay. Aarhus Harbor Bath and adjacent Beach Bath provide new ways for the public to enjoy the water in all seasons. The first harbor bath designed by BIG was built in Copenhagen in 2002 to define the Danish capital as one of the most livable cities in the world. Extending an adjacent park with cliffs, boat ramps, it was recognized by the International Olympic Committee as a 2007 Best Public Recreational Facility honorable mention.

项目名称：Aarhus Harbor Bath / 地点：Aarhus Ø, Denmark / 事务所：BIG / 合伙人负责人：Bjarke Ingels, Finn Nørkjær, Andreas Klok Pedersen / 项目负责人：Jesper Bo Jensen, Søren Martinussen / 项目团队：Annette Birthe Jensen, Franklin Natalino Simao, Giedrius Mamavicius, Jacob Lykkefold Aaen, Jakob Ohm Laursen, Johan Bergström, Kristoffer Negendahl, Lucian Tofan, Nicolas Millot, Richard Howis, Ryohei Koike, Soo Woo / 合作者：CASA A/S, CC-Design A/S, Gehl / 功能：public space / 客户：Center for Byens Anvendelse, Aarhus Kommune/Salling Fonden / 面积：2,600m² / 竣工时间：2018.6.30 / 摄影师：©Rasmus Hjortshoj (courtesy of the architect) - p.162~163, p.166~167, p.170bottom, p.171; ©Federico Covre - p.164, p.165, p.170top, p.172~173, p.174

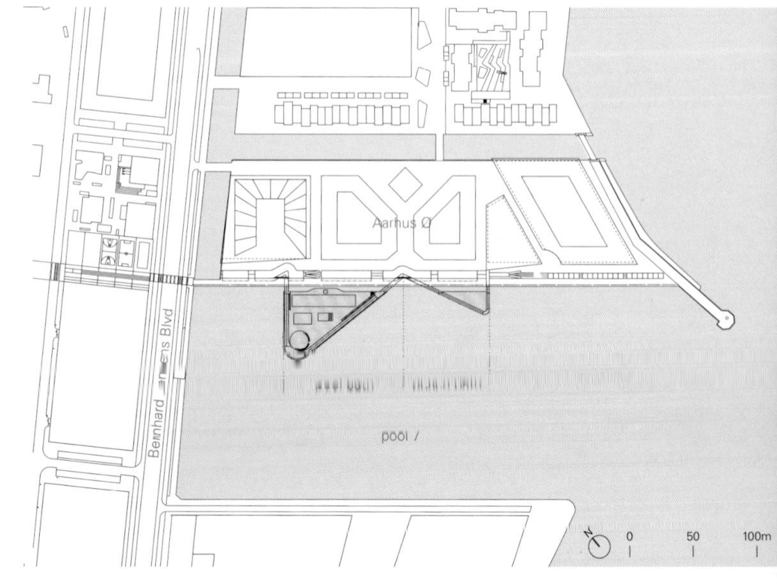

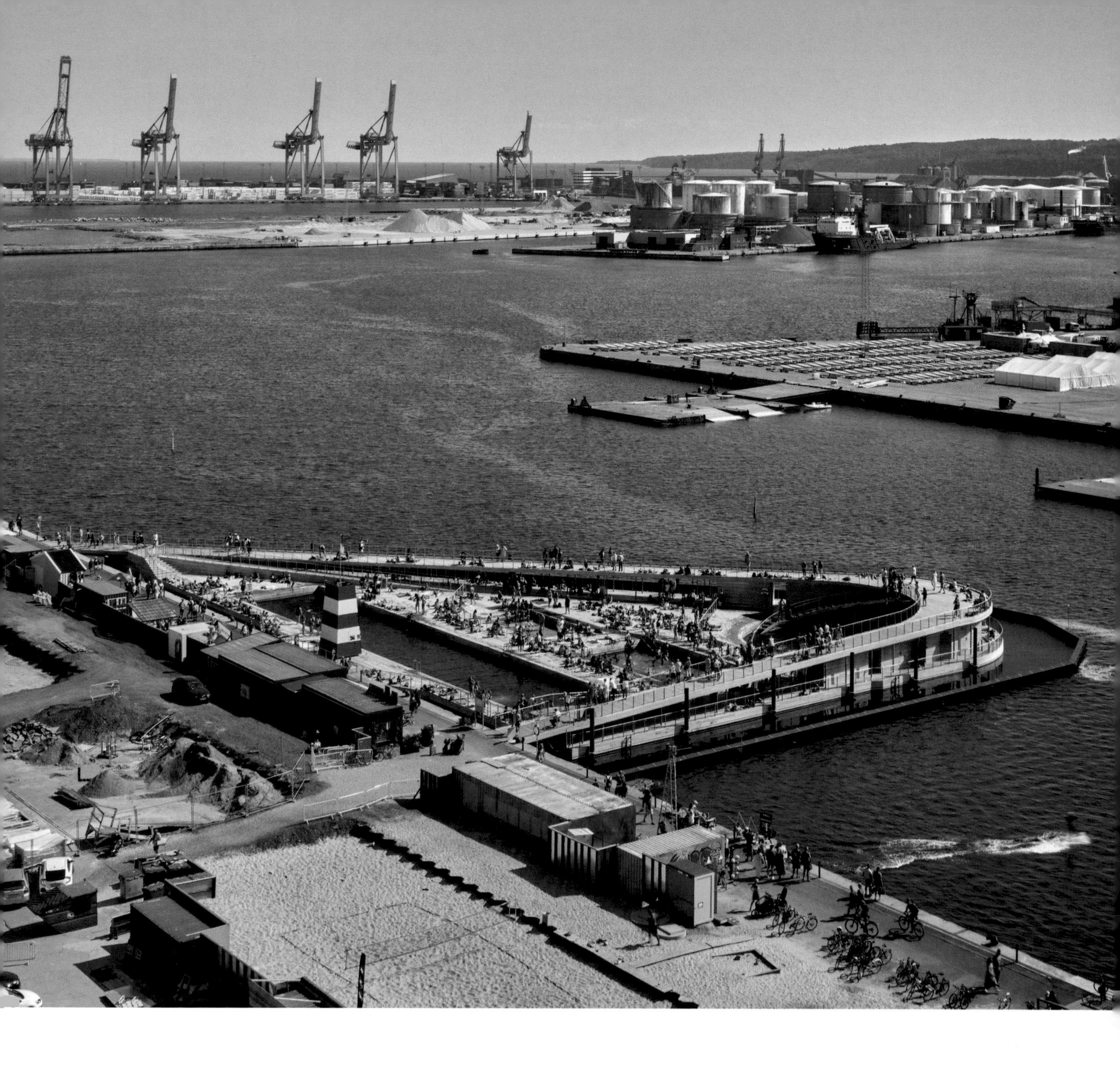

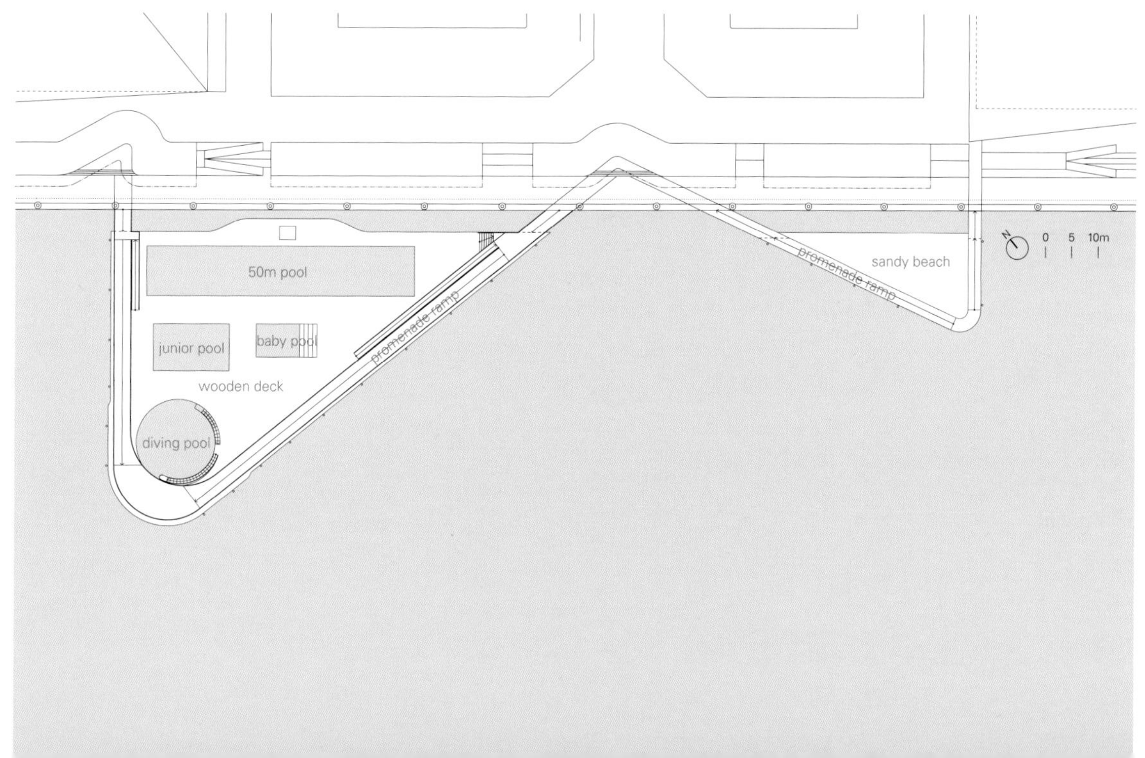

50m pool
junior pool
baby pool
wooden deck
diving pool
promenade ramp
promenade ramp
sandy beach
0
5
10m

R-E-S.eu

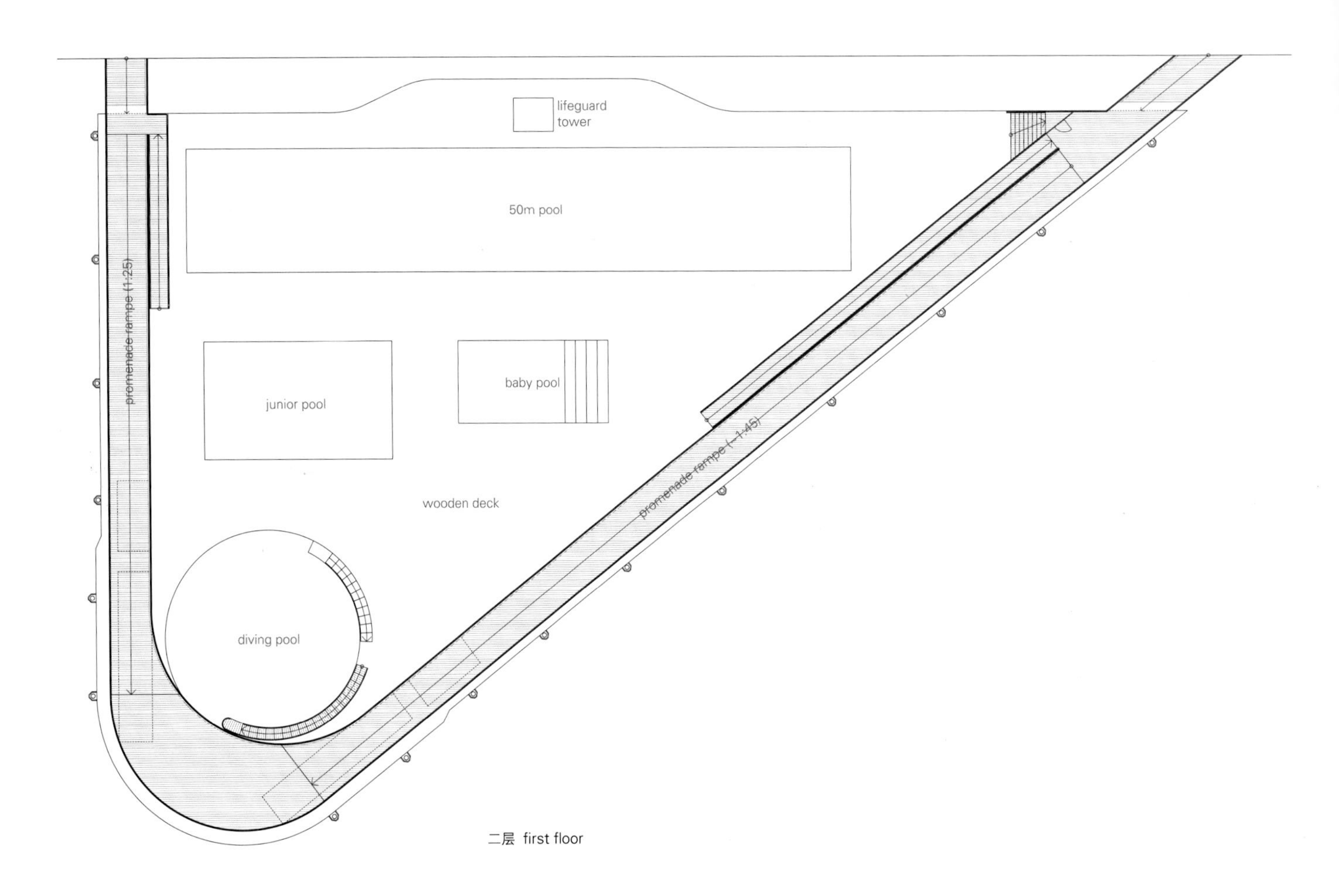

二层 first floor

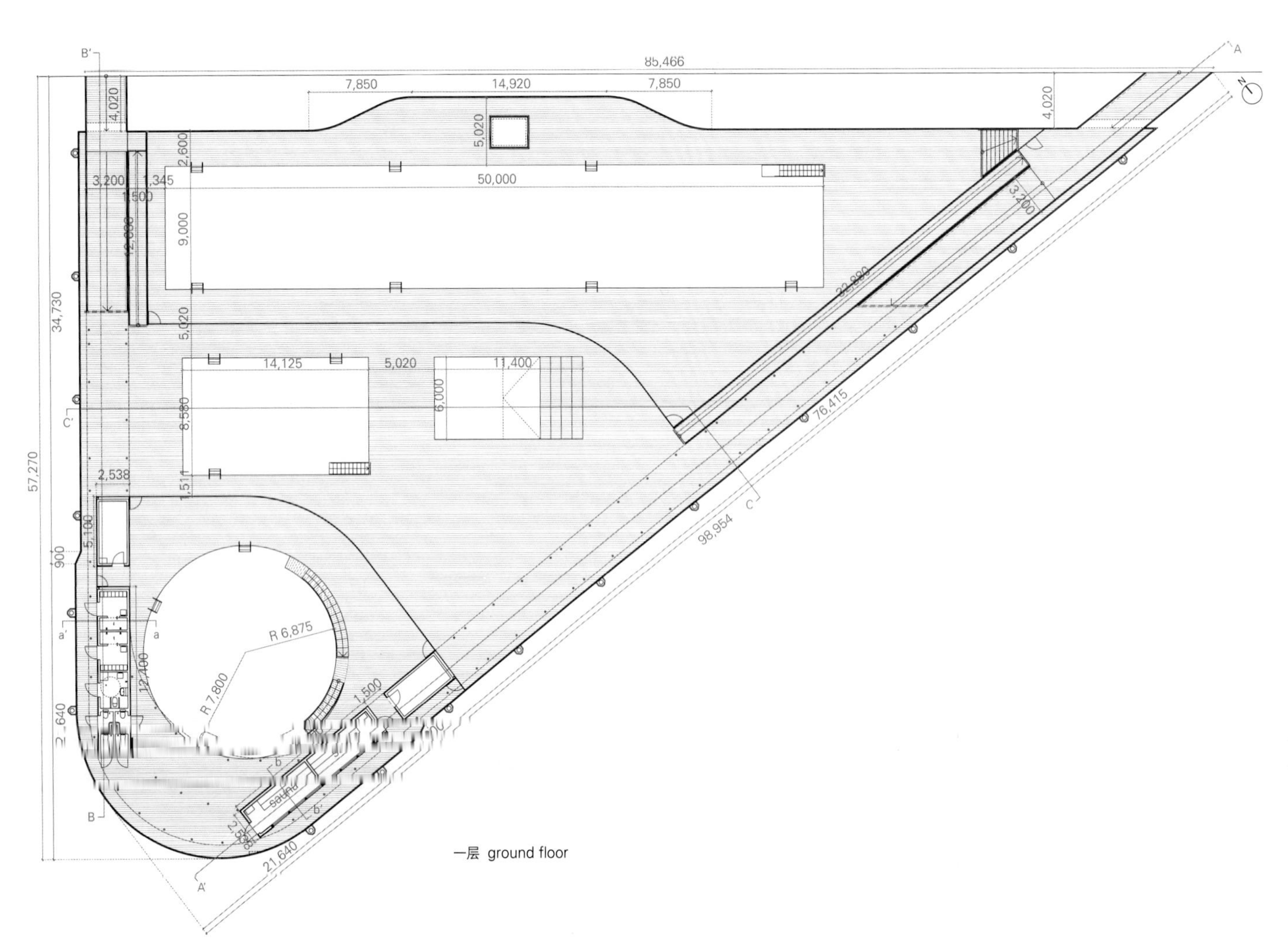

一层 ground floor

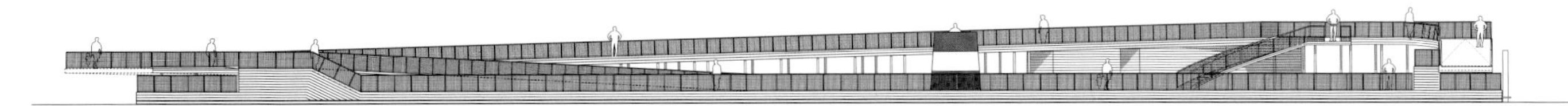

东北立面 north-east elevation

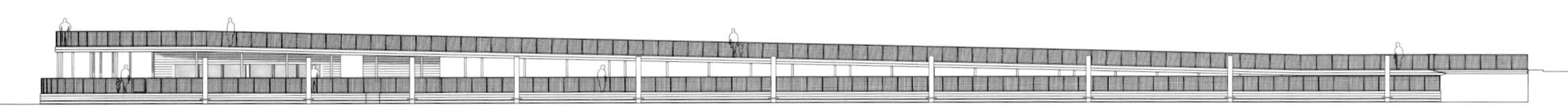

南立面 south elevation

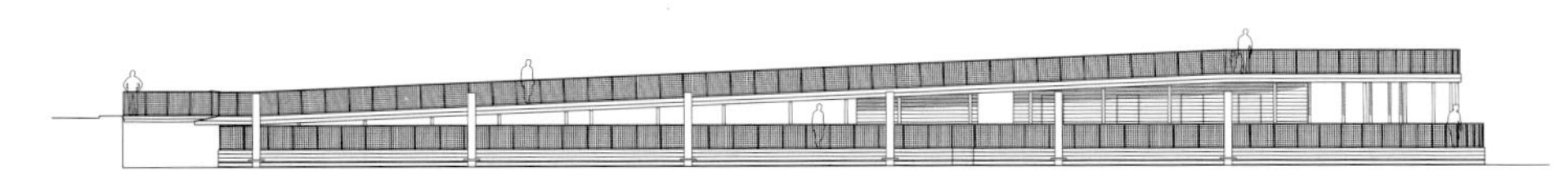

西北立面 north-west elevation

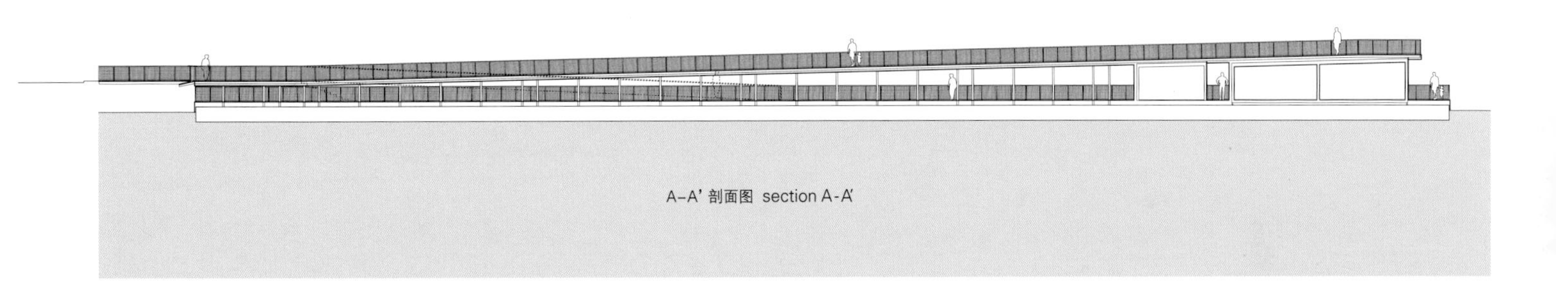

A–A' 剖面图 section A-A'

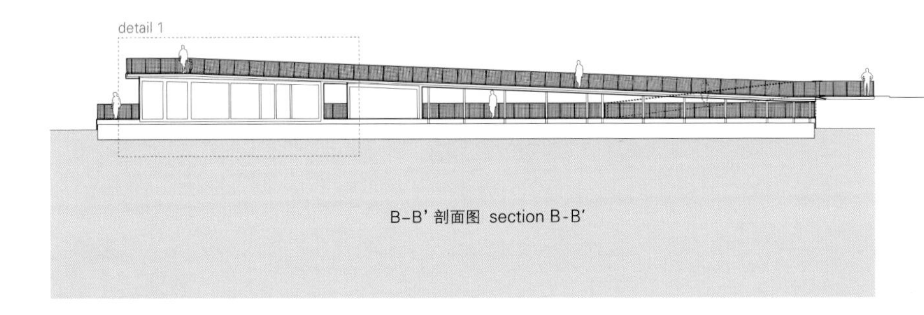

B–B' 剖面图 section B-B'

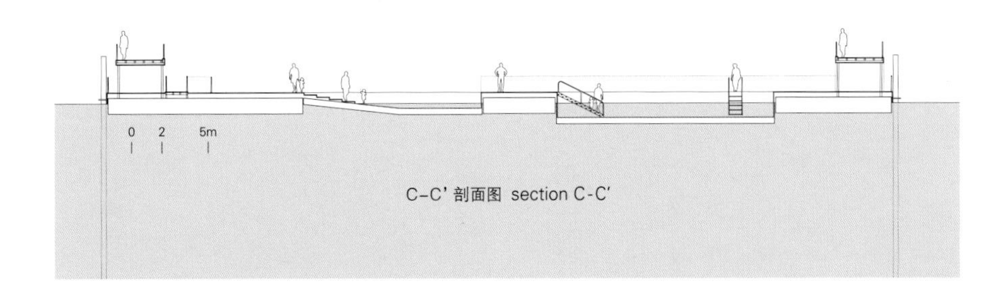

C–C' 剖面图 section C-C'

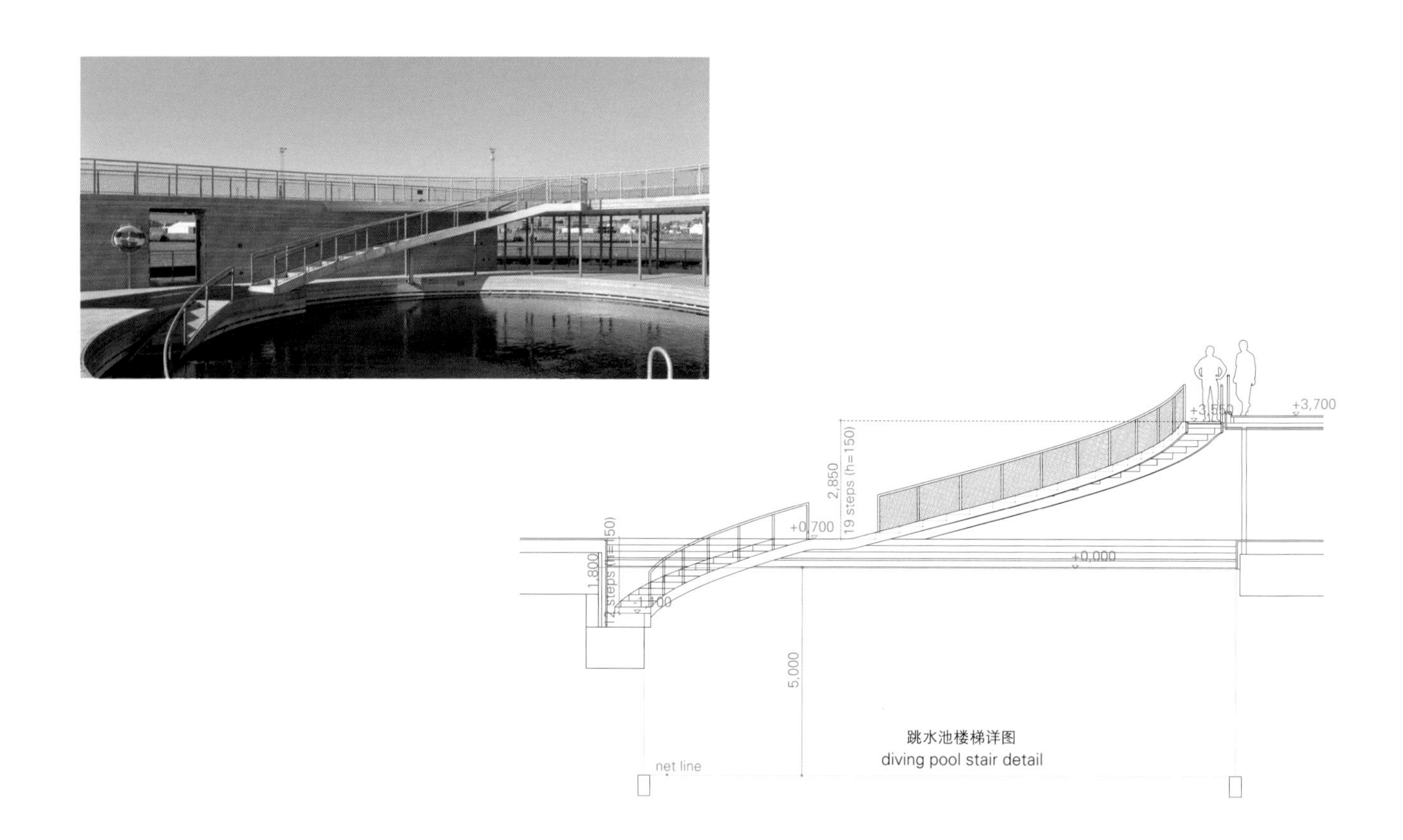

跳水池楼梯详图
diving pool stair detail

BORNHOLMSLINJEN

PAS PÅ!

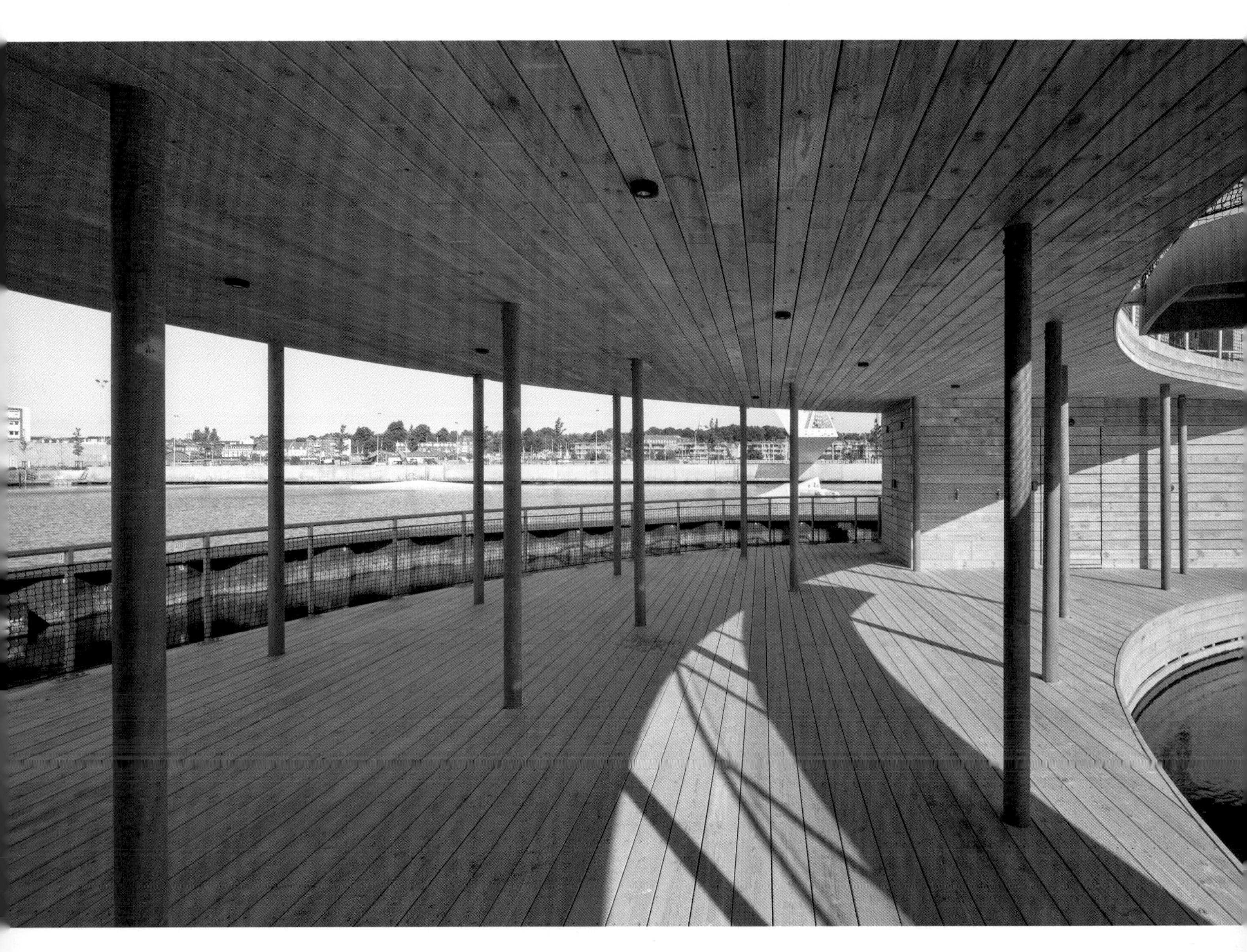

Similarly, but bigger in size and more ambitious in its design, the Aarhus Harbour Bath implemented BIG's strategy to create a framework for maximum amount of life with the minimum amount of built substance. The swimmers can enjoy the circular diving pool, the children's pool, the 50m long lap pool or one of the two saunas. Above the saunas is the public elevated boardwalk for those who prefer to stay dry, which doubles as a viewing platform overlooking the pools and water beyond.

The Harbor Bath is not just for swimmers, instead is a spot full of activities. In front of the bath, there is a series of freestanding restaurants, beach volley courts, a children's theater, beach huts for various activities and other life-creating public oriented programs. In the coming years, private building blocks including spectacular conference center called "AARhus" will rise as well. As a result, the private residential buildings with a courtyard at its center, in a range of heights at Aarhus Ø become subordinate to the needs of the public realm, altogether forming an intriguing cityscape with a recently completed public library. By designing the public space as the first step, the masterplan carefully mixes public programs with private residences, creating a new dynamic urban area where public and private realms converge.

Believed to be the largest seawater structure of its kind, it opens every weekend, accommodating 650 people, except for promenade deck which is open 24 hours throughout the year. In Denmark's second-most populous city, Aarhus Harbor Bath will give the residents and visitors of the island a more engaging and adventurous waterfront experience and breathe life into an area historically reserved for industrial purpose by claiming the water's edge as public realm.

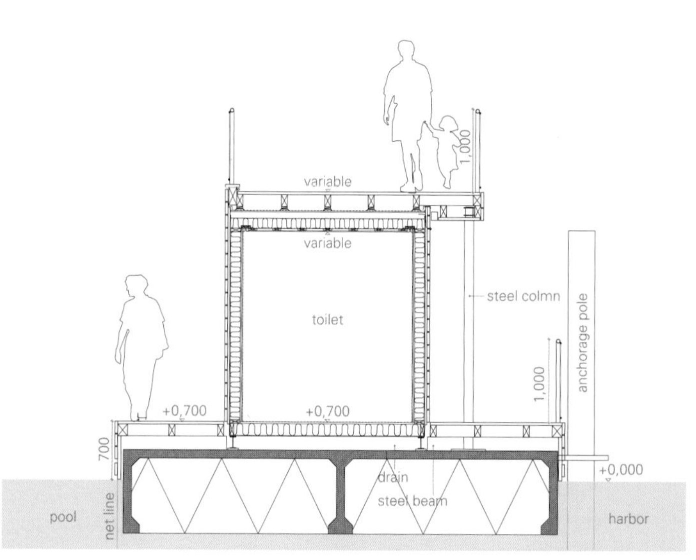

a-a' 详图 detail a-a'

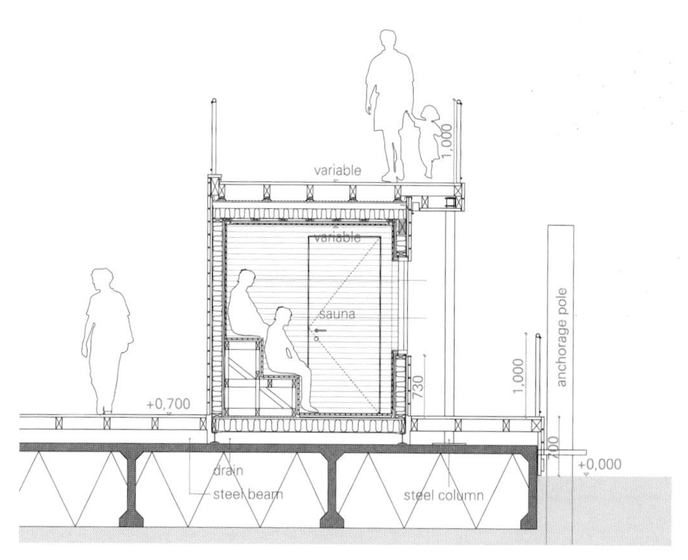

b-b' 详图 detail b-b'

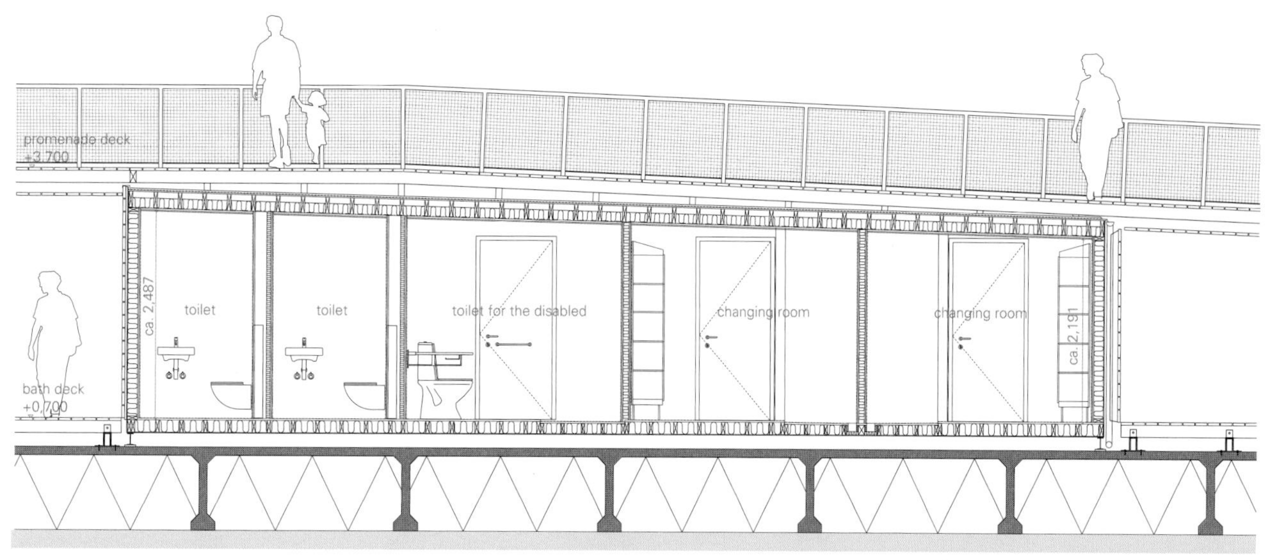

详图1 detail 1

绿湾 Titletown 公园

Green Bay Packers Titletown District

ROSSETTI

HINTER

绿湾包装工队是一支位于美国威斯康星州绿湾市的美式橄榄球球队。该球队与ROSSETTI建筑事务所合作建造了一个以体育为主题的开发项目，包括一座多功能建筑和广场，位于兰博球场东侧。

以促进绿湾市的经济发展和文化繁荣为目标，ROSSETTI制订了Titletown公园的总体规划，在不同季节和不同时间推出灵活多变的活动。设计策略更加注重公共区域的空间利用，北边广场和南边的联排别墅都有低层商业建筑。总体规划旨在更好地打造绿湾包装工队球队的品牌，刺激经济发展，开发亲子娱乐项目，使其成为一个真正具有当地特色的社区空间。

第一阶段的建造已于近期完成，已完成项目有广场、腹地啤酒厂以及艾瑞斯山。艾瑞斯山的设计相当新颖，它位于Titletown公园社区中心建筑的顶部，其下设有滑冰场。Titletown公园建筑采用的材料是对当地材料的当代诠释，如砖、钢以及鹅卵石，这些都反映了绿湾市的工业遗产。

艾瑞斯山

兰博球场的冰冻苔原在设计中被用来暗指滑雪山。建筑设计以冰冻苔原以及上面的土地为原型，艾瑞斯山被剥离，成为公园的延伸。滑雪山独特的结构设计是美国同类建筑的首例，它位于一座二层建筑的顶部，山体长度约为33.3m，下方是一个像一条冰冻的河流一样蜿蜒的滑冰场。建筑内设有餐厅/咖啡厅、零售商店、滑冰设备租赁商店以及可供租赁的活动空间。建筑材料使用的是当地的白雪松和湖底石头。以上设施能够让当地的人们在寒冷的冬季进行室外活动，而在温暖季节重新享受绿色空间与活力广场。自2017年12月正式推出冬季活动以来，新的艾瑞斯山以及滑冰场一直深受儿童和成人的喜爱。

腹地啤酒厂

腹地啤酒厂的屋顶和天窗设计体现了绿湾市的工业遗产风格。自1999年在道斯曼街开设店铺以来，腹地便成了绿湾市的标签之一。二十多年以来，屡获殊荣的啤酒厂让当地人享受着荣誉。腹地啤酒厂的位置对于Titletown公园的发展至关重要。在温暖的季节，面对一个室外天井开放的外墙可供游客通行进入与其相邻的公园。Titletown公园在美学设计和材料选取上参考了威斯康星州当地的建筑传统风格，例如，裸露钢框架和用于室内采光的天窗。就像精酿啤酒只需用几种关键成分即可酿成一样，使用少数几种简洁的建筑材料，如砖和石灰石，就能够反映这种设计理念。

第二阶段工程包括办公楼、商业区、住宅区以及大约24 280m²的公共场地。Titletown公园将成为一个有价值的、全方位的社区公园。

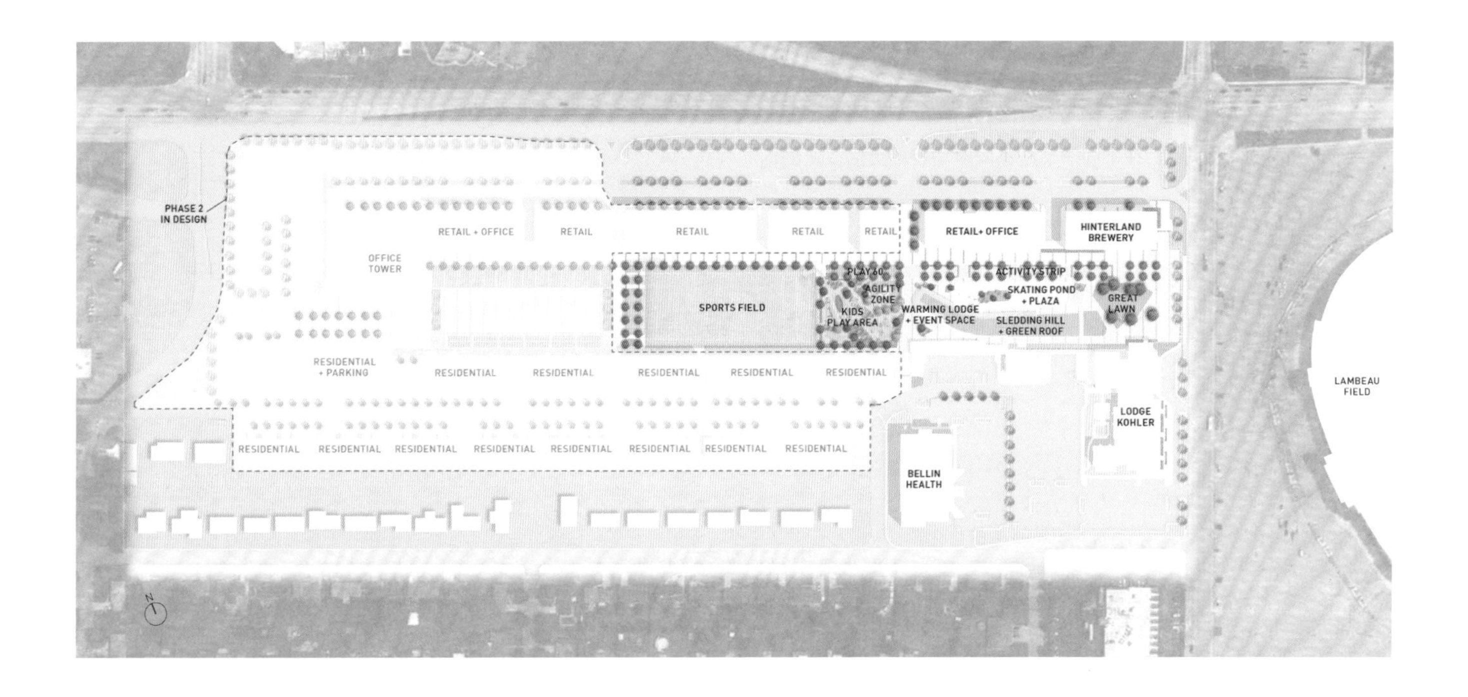
PHASE 2
IN DESIGN
RETAIL + OFFICE
RETAIL
RETAIL
RETAIL
RETAIL
RETAIL+ OFFICE
HINTERLAND
BREWERY
OFFICE
TOWER
SPORTS FIELD
PLAY 60
AGILITY
ZONE
KIDS
PLAY AREA
WARMING LODGE
+ EVENT SPACE
ACTIVITY STRIP
SKATING POND
+ PLAZA
SLEDDING HILL
+ GREEN ROOF
GREAT
LAWN
RESIDENTIAL
+ PARKING
RESIDENTIAL
RESIDENTIAL
RESIDENTIAL
RESIDENTIAL
RESIDENTIAL
RESIDENTIAL
RESIDENTIAL
RESIDENTIAL
RESIDENTIAL
RESIDENTIAL
RESIDENTIAL
RESIDENTIAL
RESIDENTIAL
LAMBEAU
FIELD
LODGE
KOHLER
BELLIN
HEALTH
N

ARIENS | HILL

The Green Bay Packers, a professional American football team based in Green Bay, partnered with ROSSETTI to bring their vision alive for a new sports-anchored development east of Lambeau Field, including a mixed-use development and plaza.

ROSSETTI developed the Titletown master plan to create a destination for Green Bay community's economic and cultural growth and promote flexible programming throughout each season and time of day. The design strategy focused on the vibrancy of the public realm with low-rise commercial buildings lining the plaza to the north and townhomes along the southern edge. The master plan was designed to enhance the Packers brand, stimulate economic activity, promote a homegrown and authentic community, and focus on family entertainment programming.

Phase 1 was recently completed with the opening of the

艾瑞斯山
Ariens Hill

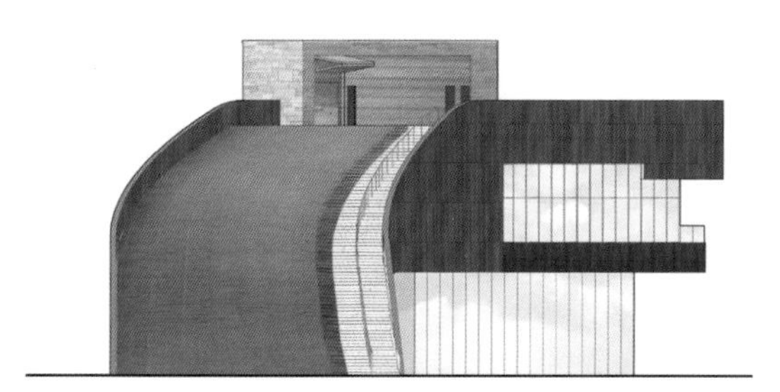

东立面 east elevation

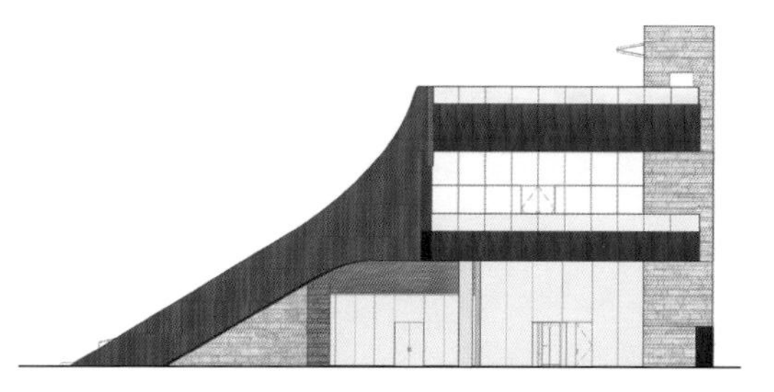

西北立面 north-west elevation

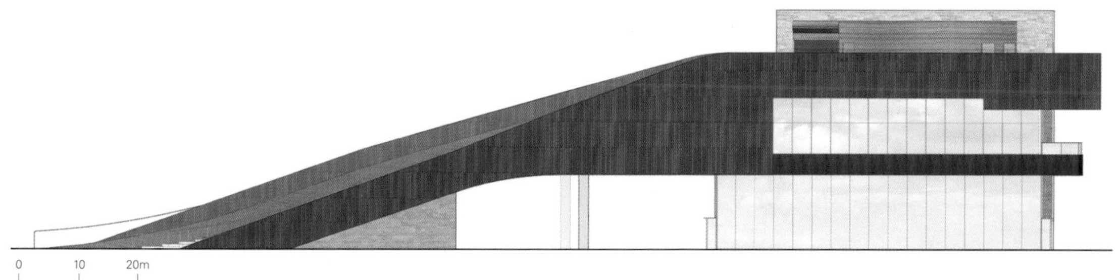

北立面 north elevation

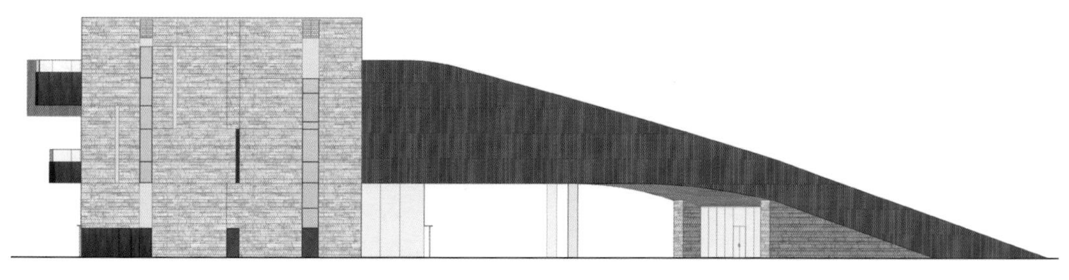

南立面 south elevation

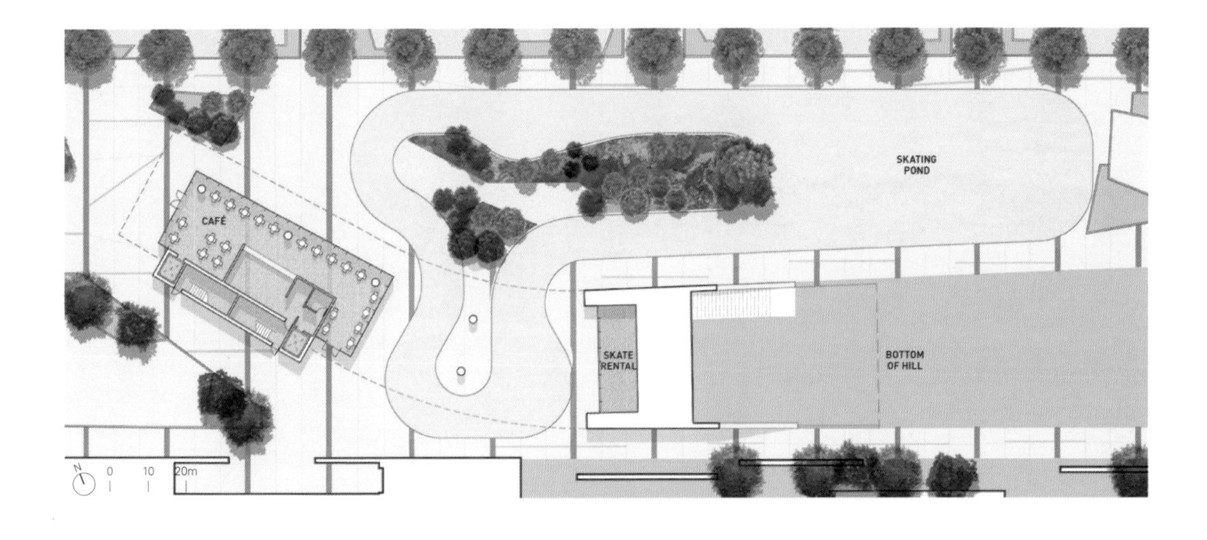
SKATING
POND
CAFÉ
SKATE
RENTAL
BOTTOM
OF HILL
N
0
10
20m

项目名称：Green Bay Packers Titletown District / 地点：Green Bay, Wisconsin, USA / 事务所：ROSSETTI / 设计团队：Jon Disbrow (design lead), Mike Shea (project manager), Dave Andruccioli, John Bigtacion, Ramon Corpuz, Deena Fox, Jhana Frederiksen, Kenton Higgins, Carla Landa, Rana Malik, Thomas Pustulka, Mike Paciero, John Page, Chris Pine, Chloe Siamof, Tessa Sodini-Marshall, Greg Sweeney, Matt Rossetti - Ariens Hill; Jon Disbrow (design lead), Mike Shea (project manager), Nick Moriarty, John Bigtacion, Ramon Corpuz, Sande Frisen, Kenton Higgins, Matt Rossetti, Patricia Dickman, Torri Smith, Tessa Sodini-Marshall, Gregory Sweeney, Daniel Wells - Hinterland / 顾问：Geiger Engineers, Illuminart Inc., JRA FoodService Construction, Strategic Energy Solutions, Design Workshop, SDI Structures, Stevens Engineers, AEI Affiliated Engineers, Peter Bosso Associates - Ariens Hill; Avitru, Strategic Energy Solutions, SDI Structures - Hinterland / 客户：Green Bay Packers, Hinterland / 造价：$125 million / 用途：1,858m² retail, 150 room hotel, 1,858m² brewery + restaurant, 2,787m² health clinic, 50 two-story townhouses, ice skating pond pavilion + sledding hill, event area / 用地面积：137,593.118m²
建筑面积：Ariens Hill 1,049m²; sloped green roof 940m²; Hinterland 1,858m² / 施工时间：2016~2018 (phase 1 - Hinterland Brewery, Ariens Hill, Lodge Kohler, The Plaza, The Pavilion) / 摄影师：©Rafael Gamo (courtesy of the architect)

腹地啤酒厂
Hinterland Brewery

东立面 east elevation

北立面 north elevation

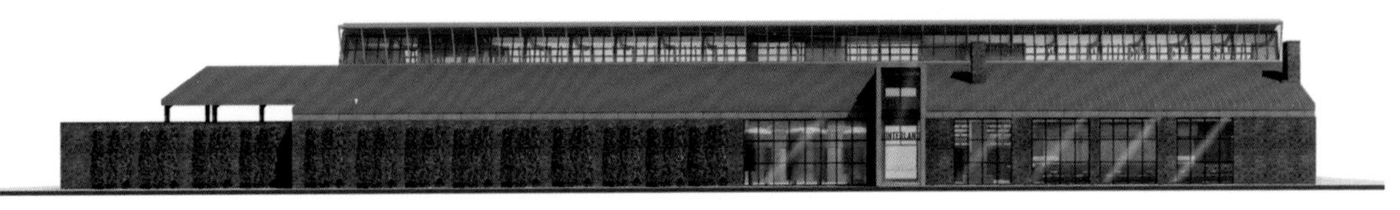

南立面 south elevation

plaza and Hinterland Brewery, as well as Ariens Hill, an innovative sledding hill designed on top of a community center with skating river below. The architecture of the district reflects the industrial heritage of Green Bay, using a contemporary interpretation of local materials, such as brick, steel and cobblestone.

Ariens Hill

Lambeau Field's Frozen Tundra was used as a design metaphor for the sledding hill pavilion. Modeling the architecture after a tundra with land above it, the hill peels up and away as an extension of the park. This unique structure is the first of its kind in the U.S. The sledding hill stretches 33.3m from start to finish and sits on top of the 2-storey pavilion with a skating pond winding underneath like a frozen river. A restaurant/café, retail shop, ice-skating rental, and leasable event space are programmed inside the space, which was built using local white cedar and Fond du Lac stone. These assets strengthen activity in the district during the winter months and have the flexibility to turn into a green space and plaza during warmer months. Both children and adults have been enjoying the new hill and skating pond since it officially opened for winter activities in December, 2017.

Hinterland Brewery

With a factory roofline and skylights, the industrial legacy of Green Bay is displayed in the Titletown home of Hinterland Brewery. Hinterland has been a Green Bay staple since they opened up shop on Dousman Street in 1999 – locals have been enjoying the award-winning brewery's libations for over two decades. Its location in Titletown is an anchor to the development, allowing patrons access to the adjacent park via an exterior wall which opens into an outdoor patio in the warmer months. Aesthetic and materials references to regional Wisconsin building traditions and sources were fundamental considerations, such as exposed steel framing and skylights for daylighting of interior spaces. Just as craft beer is brewed with only a few key ingredients, the simplicity of the building materials and minimal palette, like brick and limestone, was chosen to reflect that honesty.
With the following completion of Phase 2, comprising office building, commercial, residential along with approximately 24,280m² of Parkland, Green Bay Titletown will be a valuable, all-round community asset.

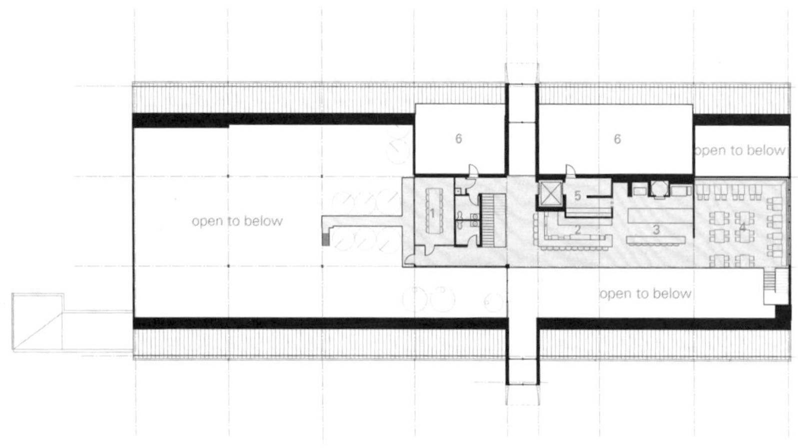

夹层 mezzanine

1. 私人餐厅
2. 酒吧
3. 厨房
4. 精致餐厅
5. 准备室
6. 机械室

1. private dining
2. bar
3. kitchen
4. fine dining
5. prep
6. mechanical room

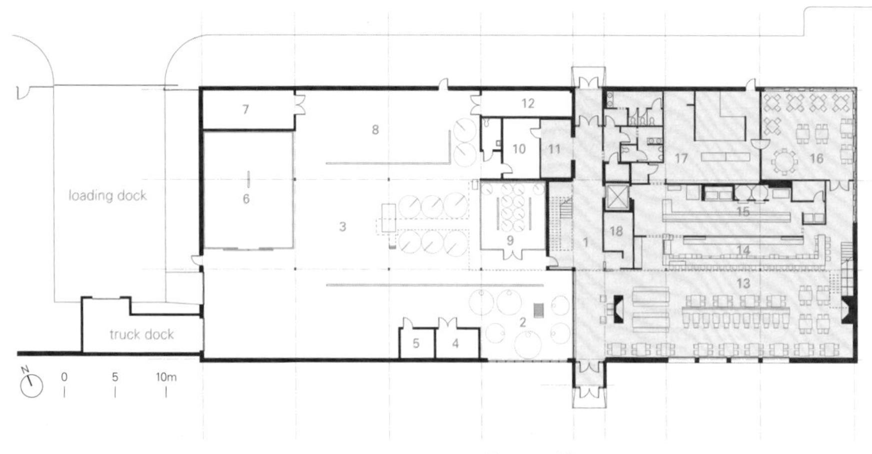

一层 ground floor

1. 大堂
2. 酿酒屋
3. 啤酒厂
4. 研磨间
5. 实验室
6. 冷藏室
7. 杂物间
8. 装瓶室
9. 冷库
10. 办公室
11. 零售空间
12. 机械室
13. 餐厅
14. 酒吧
15. 烹饪室
16. 私人餐厅
17. 准备厨房
18. 衣帽间

1. lobby
2. brew house
3. brewery
4. milling
5. lab
6. cold storage
7. utilities
8. bottling and canning
9. cold room
10. office
11. retail
12. mechanical
13. dining
14. bar
15. cooking
16. private dining
17. prep kitchen
18. coats

天理市车站广场 CoFuFun

Tenri Station Plaza CoFuFun

Nendo

日本奈良县天理市车站广场的总体规划面积为7700m²，规划建设自行车租赁区、咖啡馆、商店、信息亭、游乐区、户外舞台和会议区等区域。该项目的目标是通过为当地居民提供活动场所、旅游信息中心和休闲设施来鼓励当地社区的振兴。

天理市的周边地区有许多古老的日本陵墓，被称为"cofun"。美丽的cofun融入了城市日常的生活空间。车站广场的景观因cofun的点缀而丰富。除此之外，cofun还体现出了该地区地理地貌的特征，即奈良盆地是四面环山的。

在广场上建造圆形cofun结构的建筑技术是将一块块披萨一样的巨型混凝土预制模具装配在一起。因为混凝土预制模具首先是在工厂成型，然后在现场组装的，因此施工结构的精确性得到了保证。同时由于同一模具可以多次使用，因此它的性价比极高。预成型的部件像建筑一样组装在一起，使用的是用来建造桥梁的巨型起重机，可以在不使用柱或梁等结构的情况下创造一个大型的空间，并且圆形形状为建筑的良好平衡提供了稳定性，以抵抗来自任何方向的力。

不同高度的cofun功能不同：有的cofun是楼梯和长凳供人坐下休息，有的confun是围栏可以保护玩耍的孩子，有的cofun是咖啡馆和舞台屋顶，有的confun是用于展示商品的架子，有的cofun则用于夜间照明，这些光亮让整个广场明亮起来。用途多样的cofun可以使人们在广场中的不同地方探索并消遣时光，而不是将游客限制在一个地方活动。cofun是一个"模糊"的空间，它既是咖啡馆、游乐场，也是一件巨大的家具。广场的路标和广告牌都有着和cofun一样的柔和曲线，深灰色的颜色既保证了醒目又能与周围环境完美融合。为了减少噪声，它们根据功能的不同，按照四种不同的高度排列。儿童游乐区、阅读书籍的休息室学习区以及可用于举办音乐会或进行公共放映的舞台都设在会议区域。除此之外，人们还可以在会议区域旁边的一家新设计的商店内购买天理市纪念品。

广场每一处的设计都尽量让cofun内部和外面广场的材料和颜色相匹配。cofun的内部陈设和固定装置都是由奈良县木材制成的，设计围绕着cofun主题，与整个广场的风格十分和谐。

作为广场的名字，CoFuFun将设计主题"cofun"和日本口语表达相结合。"Fufun"的意思是开心，指的是快乐和无意识的嗡嗡声；广场的

设计要提供一种欢乐的气氛，让游客在参观时不自觉地愉快哼唱。

"CoFuFun" 也寓意 "cooperation（合作）" 和 "community（社区）" 中的 "co-" 以及 "fun（乐趣）" 本身。虽然本身是日语和英文字母的拼写，但却有着相似的意义，可以让外国游客也轻松、容易地理解。

The 7,700m² area master plan for the station plaza at Tenri Station in Nara Prefecture, Japan includes bicycle rentals, a cafe and other shops, an information kiosk, a play area, an outdoor stage, and a meeting area. The project goal was to encourage local community revitalisation by providing a space for events, tourist information center and leisure facilities for local residents.

Tenri's urban boundaries consist of a number of ancient Japanese tombs, known as "cofun". The cofun, beautiful and unmistakeable, blends into the spaces of everyday life in the city. The plaza's landscape, richly punctuated by several of these cofun, is a representation of the area's characteristic geography: the Nara Basin, surrounded on all sides by mountains.

The construction technique used to create the plaza's round cofun structures consisted of fitting together pieces of a precast concrete mould resembling a huge pizza. Because precast concrete moulds are formed at the factory and then assembled on site, the resulting structures are precise. Also since the same mould can be used multiple times, it achievd excellent cost-performance. The pre-formed parts are pieced together like building blocks using the same

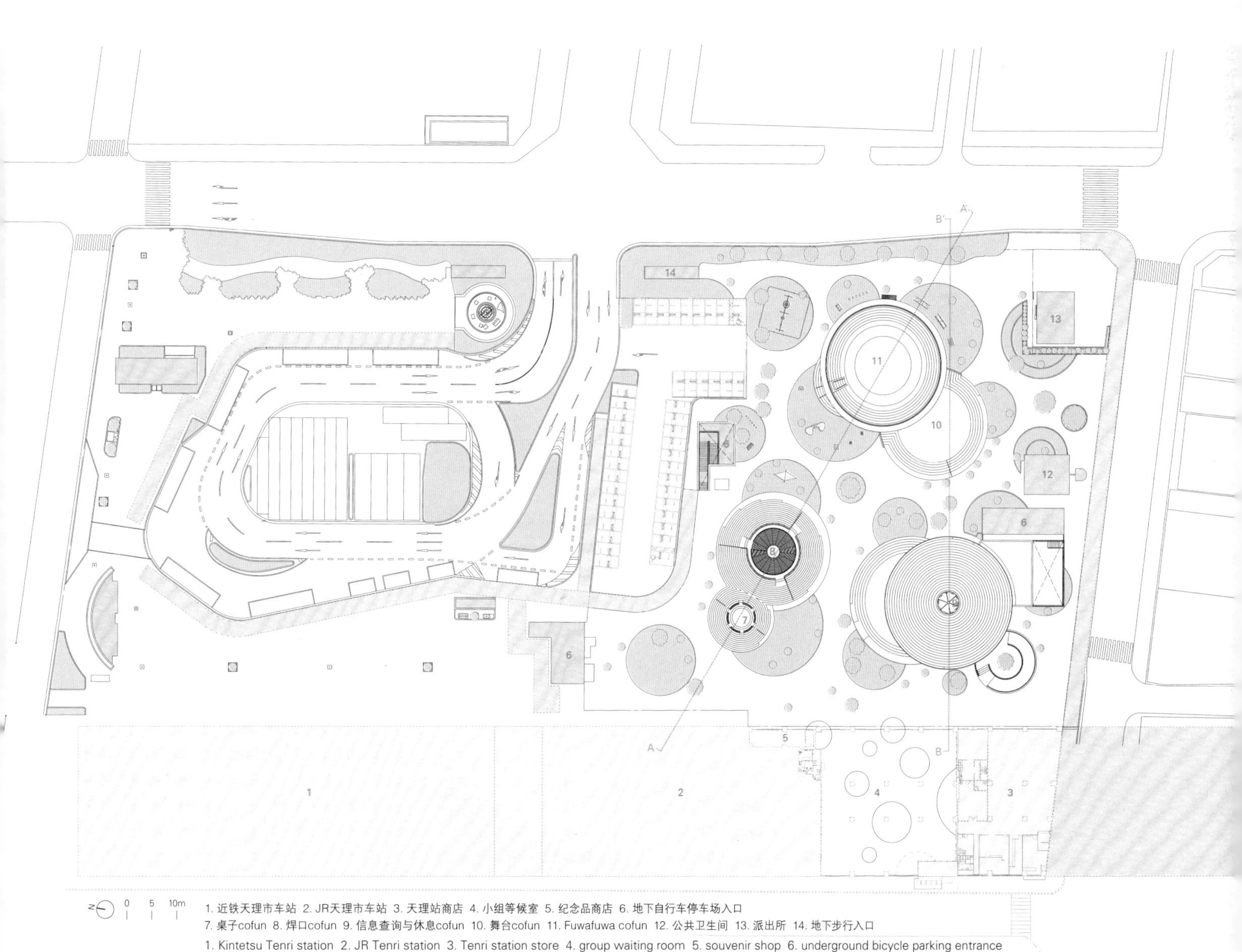

1. 近铁天理市车站 2. JR天理市车站 3. 天理站商店 4. 小组等候室 5. 纪念品商店 6. 地下自行车停车场入口 7. 桌子cofun 8. 焊口cofun 9. 信息查询与休息cofun 10. 舞台cofun 11. Fuwafuwa cofun 12. 公共卫生间 13. 派出所 14. 地下步行入口

1. Kintetsu Tenri station 2. JR Tenri station 3. Tenri station store 4. group waiting room 5. souvenir shop 6. underground bicycle parking entrance 7. Table cofun 8. Crator cofun 9. Info & Lounge cofun 10. Stage cofun 11. Fuwafuwa cofun 12. public lavatory 13. Koban 14. underground walking entrance

massive cranes used to build bridges. Large spaces can be formed without the use of columns or beams, and the round shape offers stabilitiy for the well-balanced structures against forces applied from any direction.
The cofun's different levels serve a variety of purposes: they are stairs and benches for sitting, fences to protect playing children, the cafe and stage roofs, shelves for displaying products and the nighttime lighting, which floods the plaza with light. This variety creates an environment that encourages visitors to explore and spend time in different spaces within the plaza, rather than limiting their movement to one place. It's an "ambiguous" space that is entirely a cafe, a playground and a massive piece of furniture, all at once.
Guideposts and signboards feature gentle curves similar to those of the cofun, and are colored in dark grey that creates a natural contrast while fitting in with the surrounding area. They are also arranged at four different heights according to their function in order to minimize noise levels. A play space for children, a lounge and study space for reading books, and a stage that can be used for concerts or public screen-

ings have all been added to the meeting area, and Tenri souvenirs can be purchased at a newly designed shop next to the space.

Every design was given to ensure that the materials and coloring of the interiors matched those of the plaza as closely as possible. Furniture and fixtures are made of wood from Nara prefecture and designed around a cofun theme to create a sense of uniformity with the plaza.

The plaza's name, CoFuFun, combines the main design motif, the cofun, with colloquial Japanese expressions. "Fufun" refers to happy, unconscious humming: the design for the plaza should offer a convivial atmosphere that unconsciously leads visitors to hum happily while they visit.

The spelling, "CoFuFun", also brings in "co-" of "cooperation" and "community", as well as, of course, "fun" itself. The result is a name whose Japanese and English spellings mean similar things, so that foreign visitors will understand it in the same way too.

©Daici Ano (courtesy of the architect)

©Daici Ano (courtesy of the architect)

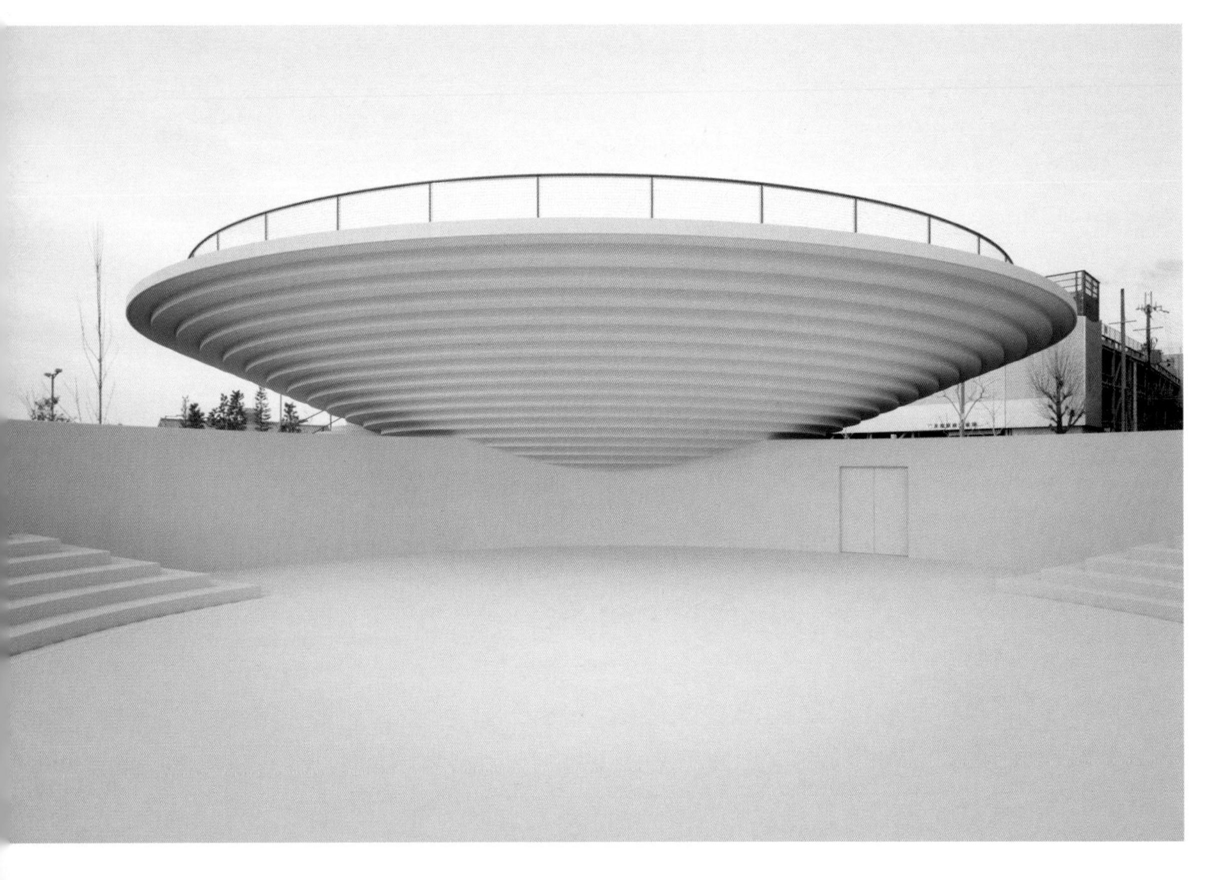

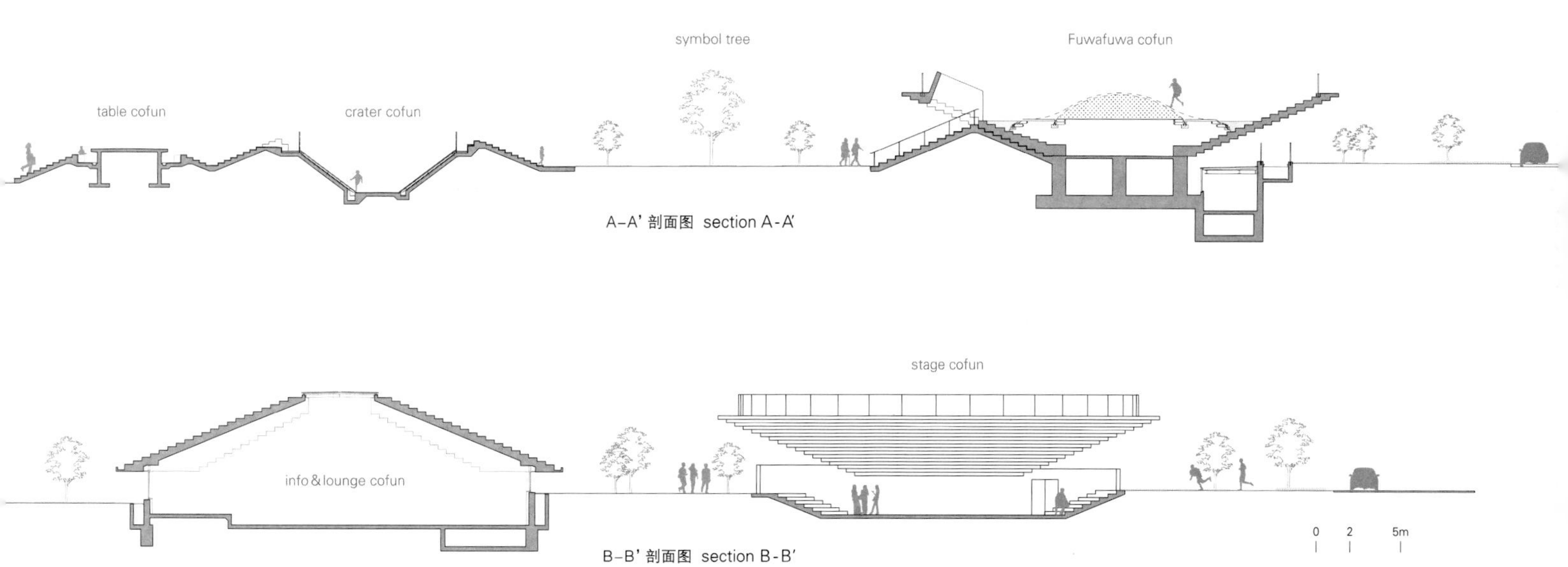

A–A' 剖面图 section A-A'

B–B' 剖面图 section B-B'

项目名称：Tenri Station Plaza CoFuFun / 地点：Tenri City, Nara, Japan / 事务所：nendo / 合作者：awn, oni, vac, Iwataya Architects, Nippon Design Center Irobe Design Institute, Izumi Okayasu lighting design, Studio Mons / 结构工程师：Seed consultant Inc. / 电气工程师：Meikoh-denki corporation. / 照明工程师：Izumi okayasu / 景观建筑师：nendo, Iwataya Architects / 承包商：Daiwa House Industry Company, Limited and Okatoku construction Company, Limited / 客户：Tenri city / 用途：public area, park / 用地面积：17,144.45m² / 建筑面积：1,130.3m² / 总楼面面积：7,717.4m² / 结构：precast prestressed concrete and steel construction(part of) / 室外饰面：polymer modified cement waterproofing-roof; acrylic resin paint-wall; natural color pavement-outer structure; aluminum sash fluoropolymer baking finish-outer fittings / 室内饰面：concrete gold trowel finish-floor; plaster board + synthetic resin emulsion paint-wall; board + synthetic resin emulsion gross paint-ceiling / 造价：1,000,000,000 JPY
竣工时间：2017.4 / 摄影师：©Takumi Ota (courtesy of the architect) (except as noted)

Park Side Kitchen
LOCAL FOODS

山湖公园游乐场

Bohlin Cywinski Jackson

山湖公园游乐场坐落于旧金山里士满区，位于一排爱德华式住宅的后面，它依偎在成熟的常绿乔木和山区湖畔的斜坡中（缓缓延伸向普雷西迪奥国家公园南缘）。翻新后的游乐场与周围景观融为一体，给人以这样的感受：这个新的游乐场本来就在这里，让人很熟悉，仿佛它本来就属于这里，属于旧金山的这个区域。

翻新后的游乐场利用地形，按照孩子的年龄和设施的可玩性在阶梯上设有单独的游乐区，游乐区通过一系列蜿蜒的小路连接在一起，这些小路为整个游乐场提供通往滑梯顶部的通道。游乐场的核心部分是一个珍贵的混凝土滑梯，它从一个巨大的土堆一侧倾泻而下，这一珍贵部分的保留，将翻新的游乐场定格在社区过去的珍贵回忆中。在通往滑梯顶部的途中会经过一个观测平台，它坐落在一片钢柱森林中，让人不禁想起周围公园的树木。人们通过这个平台不但可以俯瞰下面的幼儿游乐区，同时还可以看到广阔的山湖景观。透过平台上的圆形孔洞，上面和下面的游客有时能够看到彼此，乐趣横生。

其他的设计元素借鉴了该地区丰富的自然历史。学前教育区的"沙丘"代表了曾经遍布整个地区的滚滚沙丘；混凝土墙的棱纹图案来自山湖水边芦苇的抽象图案。原生于此地的鸟类和动物的足迹被印在与学龄区紧挨的墙壁上；像青蛙和海龟这样的大型雕塑，则代表湖中原生的水生生物。大型的木质游乐设施给人们以这样的印象——它们是用周围森林里的原木制作而成的，进一步体现了就地取材的特色。

翻修游乐场的工作由当地的三位母亲带头，旧金山娱乐部门和公园部门参与了合作。这三名女性在了解到游乐场的翻新只有在社区倡导者的支持下才能进行之后，于2010年成立了"山湖公园游乐场之友"组织。经过山湖公园游乐场之友组织的初步努力，该项目通过清洁与安全邻里债券募得了资金，其中还包括该组织本身的大量捐赠。其余资金主要来自社区的中小型捐款，以及博林·西温斯基·杰克逊建筑事务所、卢茨克协会和霍姆斯结构事务所的捐赠。

Tucked behind a row of Edwardian homes in San Francisco's Richmond District, Mountain Lake Park Playground is nestled amidst mature evergreen trees and gentle slopes down to the shore of Mountain Lake on the southern edge of Presidio National Park. The new design works seamlessly within this context, creating a feeling that the playground has always been there – specific to this place and this part of San Francisco.

The renovated playground takes advantage of the site's topography, with separate play areas on terraces, organized according to age and playability, threaded together by a series of meandering pathways that provide an accessible route throughout the playground, to the top of the slide. The centerpiece of the playground – a treasured concrete slide that cascades down the side of a large earthen mound – has been preserved, anchoring the updated playground to the cherished memories of the community's past. Midway on the journey to the top of the slide, an observation platform sits on a forest of steel columns, evocative of the trees in the surrounding park. The platform overlooks the preschool area below while facilitating expansive views of Mountain Lake. Apertures in the platform provide framed moments of intrigue for the users both above and below.

Additional design elements draw on the rich natural history of the site. The "sand dunes" of the preschool area represent the rolling sand dunes that once spread across the region; the ribbed pattern of the concrete walls is an abstraction of tulle reeds that line the shores of Mountain Lake; tracks of birds and animals native to the area imprint the surface of the wall that borders the school age area; while large sculptures, including a frog and turtle, acknowledge the native aquatic life in the lake. These site-specific references are enriched by large timber play structures, giving the impression that they were fashioned from logs of the surrounding forests.

The effort to renovate the playground was spearheaded by a group of three local mothers, in partnership with the San Francisco Recreation and Parks Department. The three women formed the "Friends of Mountain Lake Park Playground" (FMLPP) in 2010, after they learned a renovation could only happen with the support of community advocates. The project was earmarked to receive funding through the Clean & Safe Neighborhood Parks Bond because of FMLPP's initial efforts, which also included a substantial donation from the group itself. Remaining funds were raised mainly through small- and medium-sized donations from the community, as well as donated services from Bohlin Cywinski Jackson, Lutsko Associates, and Holmes Structures.

A'
C
B'
B
b'
b
a'
a
C'
A
N
0
2
5m

项目名称：Mountain Lake Park Playground
地点：San Francisco, California, USA
事务所：Bohlin Cywinski Jackson
项目团队：Gregory Mottola-principal;
Aaron Gomez-project manager;
Laing Chung, Lauren Ross, Daniel Yoder
景观建筑师：Lutsko Associates
结构工程师：Holmes Structures
土木工程师：Lea&Braze Engineering
娱乐设施：Richter Spielgeräte GmbH
with Architectural Playground Equipment
现浇橡胶表面：Robertson Recreational Surfaces
场地家具：Maglin Site Furniture
客户：Friends of Mountain Lake Park Playground,
San Francisco Parks Alliance, City and County of San Francisco
用地面积：2,043m²
竣工时间：2017
摄影师：©Nic Lehoux (courtesy of the architect)

A-A' 剖面图 section A-A'

B-B' 剖面图 section B-B'

C-C' 剖面图 section C-C'

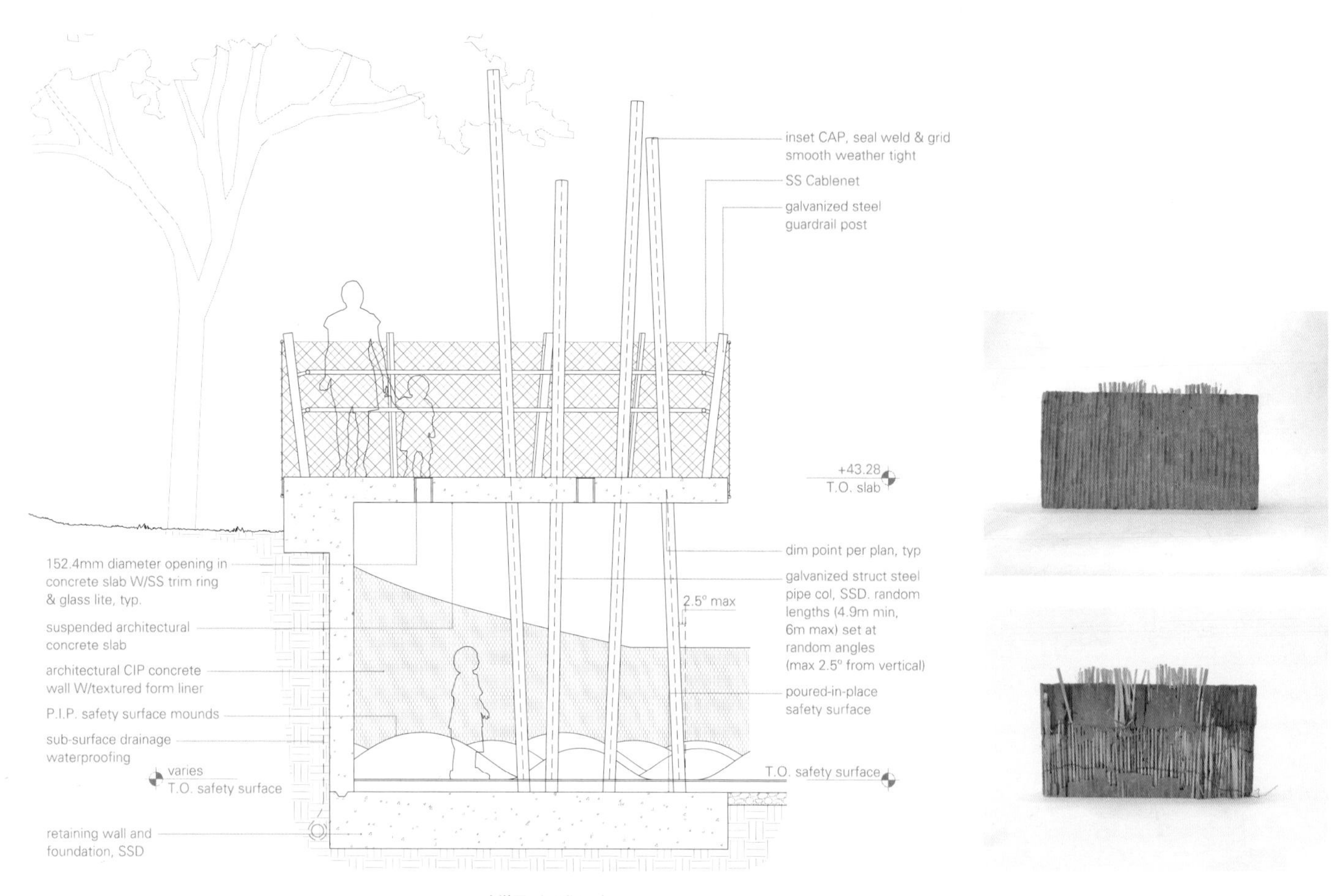

a–a' 详图 detail a-a'

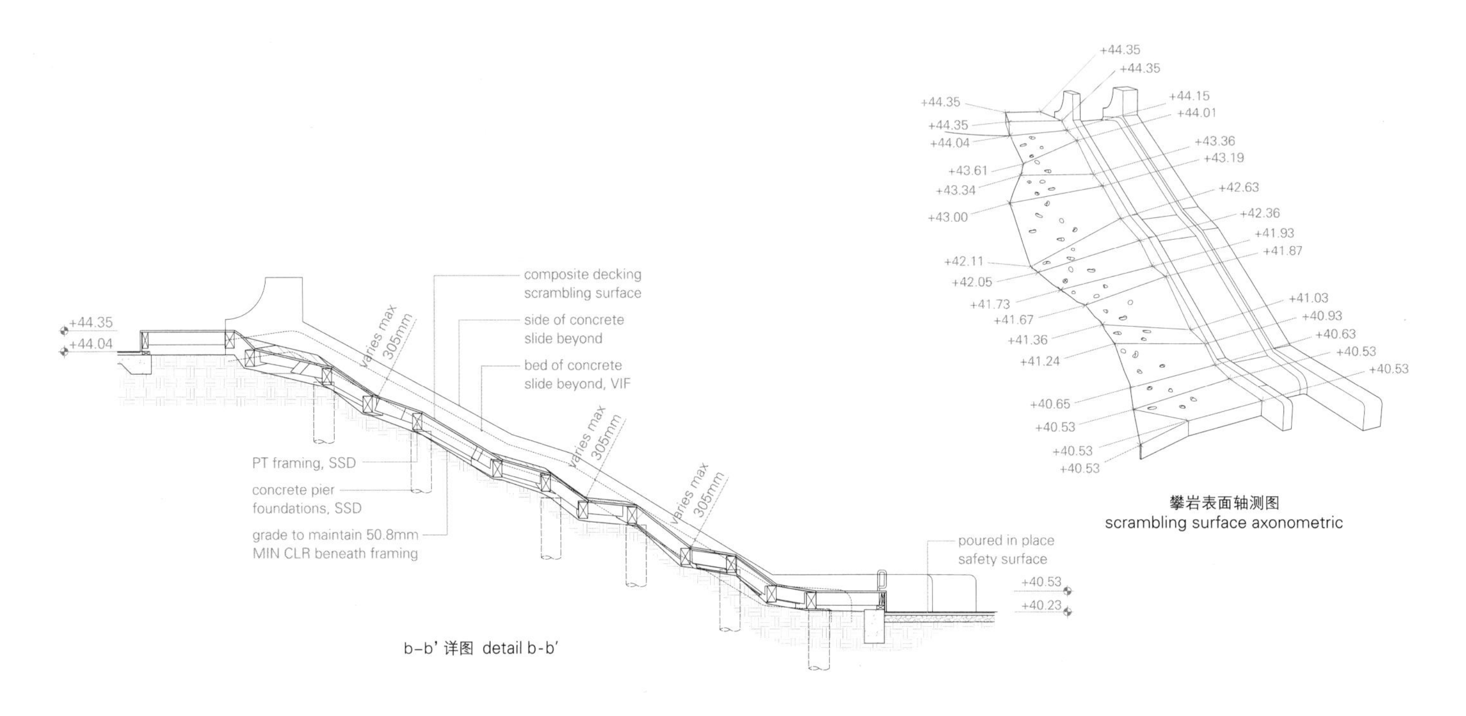

b-b' 详图 detail b-b'

攀岩表面轴测图
scrambling surface axonometric

©Tuomas Uusheimo

P12 STUDIO V Architecture

Was founded by Jay Valgora in April 2006, Manhattan. Now has 30 staff members in the office. Became known for its creative and diverse design strategies, and intensive research that reconciles urban form with contemporary architecture. Has excelled in large-scale urban design and master planning projects, ranging from new transit-oriented developments, green residential communities, public plazas, and networks of waterfront parks. Over twenty waterfront projects in the New York metropolitan area and internationally are re-defining edge design and resiliency.

P12 S9 Architecture

Founding principal, John Clifford is a licensed engineer of New York. Received his Bachelor of Civil Engineering from the Georgia Institute of Technology. With more than 25 years of design and development experience, John offers unparalleled expertise on retail, mixed-use, master planning, and office projects. Co-founder and a design principal, Navid Maqami is a Registered Architect of New York. Received his AA Diploma from the Architectural Association, London. With over 30 years of experience, Navid leads teams of designers and architects on diverse projects, from large scale mixed-use developments to small interiors. With a design approach rooted in modern contextualism and inspired by urban narratives, their works include residential, mixed and adaptive reuse projects. Among them, 205 Water Street in Brooklyn (2012) and Ponce City Market in Atlanta (2015). Currently under construction are 606 Broadway (New York) and Dock 72 (New York), all expected to be completed by 2019.

P60 JKMM Architects

Is a 100-strong group of architects and designers based in Helsinki, Finland. Has been working on major public buildings in Finland for twenty years. JKMM's work is diverse ranging from housing to schools, libraries to concert halls, museums to hospitals. Founding partner and principal architect, Asmo Jaaksi[right] (1966) graduated in Architecture from the Tampere University of Technology in 1997. Project architect, Freja Ståhlberg-Aalto[left] (1973) graduated in Architecture from Helsinki University of Technology in 2001. She has held a part time teaching position at the Department of Architecture, Aalto University. Restoration architect, Katja Savolainen[third-left] (1969) graduated in Architecture from Helsinki University of Technology in 1999. She joined JKMM in 2004. Interior architect, Päivi Meuronen[second-left] (1967) graduated in Interior Architecture from the University of Art and Design, Helsinki and has been the driving force behind JKMM Architects' interior designs since 2003.

P80 3XN

Danish architect, Kim Herforth Nielsen is founding partner, principal and creative director of 3XN since 1986. Graduated from the Aarhus School of Architecture in 1981. Member of the Danish Architectural Association, RIBA and AIA. Jury Member of the World Architecture Festival and AR Awards for Emerging Architecture. Received numerous awards including RIBA Award 2005, 2007, 2009, 2011 and 2013. Is also Honorary Professor of Aarhus School of Architecture and Chairman of architecture committee at Danish Arts Foundation. Involved in all the major projects of the practice, including the Copenhagen Arena, Blue Planet Aquarium, Museum of Liverpool, Ørestad College and the UN City HQ in Copenhagen.

P96 Heneghan Peng Architects

Is a design partnership practicing architecture, landscape and urban design. Was founded by Shih-Fu Peng and Róisín Heneghan in New York in 1999 and was relocated to Dublin in 2001. Opened their Berlin office in 2011. Have collaborated with many leading designers and engineers on a range of projects which include larger scale urban master plans, bridges, landscapes and buildings. Major projects include the Canadian Canoe Museum, the National Gallery of Ireland, the Palestinian Museum, Grand Egyptian Museum at the Pyramids, the Giant's Causeway Visitors' Centre, Airbnb EMEA headquarters and the Diamond Bridges at the 2012 London Olympic Park. Recently received RIAI President's Award for the Palestinian Museum in 2017.

P80 **HKS**

Was founded in 1939 and has 24 offices worldwide. To those wanting to make an impact with design, HKS is the company where focused people consistently deliver the extraordinary. They are creating exceptional spaces across the globe through their connected network of offices and people. With tremendous talent across a wide spectrum of expertise, they believe that the value of talent, experience and knowledge is multiplied when shared.

Diego Terna

Received a degree in architecture from the Politecnico di Milano and has worked for Stefano Boeri and Italo Rota. Has been working as critic and collaborating with several international magazines and webzines as editor of architecture sections. In 2012, he founded an architectural office, Quinzii Terna together with his partner Chiara Quinzii. Currently is an assistant professor of Politecnico di Milano and runs his personal blog L'architettura immaginata (diegoterna.wordpress.com).

P188 **Nendo**

Oki Sato was born in Toronto, Canada in 1977. In 2002, he received his M.A. in Architecture from Waseda University, Tokyo and established Tokyo office of Nendo. Afterward, he established his Milan office in 2005. Was a jury member of iF award in 2010 and a lecturer at the Waseda University, Tokyo in 2012. In 2015, he was selected as Designer of the Year by Maison & Objet and Interior Designers of the Year by Iconic Awards. Was a producer of World Design Capital Taipei and Jury Chair for the Golden Pin Design Award in 2016.

P200 **Bohlin Cywinski Jackson**

Is an American architecture firm founded in 1965 by Peter Bohlin and Richard Powell. 13 principals including Gregory R. Mottola, and staff of practice architecture, planning and interior design work. Associate, Aaron Gomez joined the San Francisco office in 2006. Received his B.A. Arch from University of New Mexico and M. Arch from Columbia University in New York. Is a registered architect of New Mexico and American Institute of Architects. Believes every project deserves to be as unique as the circumstances that surround it; that a building is tailored to those circumstances by considered editing and thoughtful addition.

P134 **Richter Architects**

Was established in 1993 by David Richter[right] and his partner Elizabeth Chu Richter[left]. Their projects, ranging from governmental, educational, religious, commercial, and residential, have been widely published and has received more than 50 design awards. Named as The Firm of the Year by the Texas Society of Architects in 2011. David Richter received his B.Arch with Magna Cum Laude from the University of Texas in 1974. Was elected 1998 President of the Texas Society of Architects. Served as a Trustee for the Texas Architectural Foundation and as President of the Foundation from 2006~2010. Elizabeth Chu Richter received her B.Arch from the University of Texas at Austin. Was elected to serve as the National President of the AIA in 2015. Is a recipient of the President's Medal from the Royal Architectural Institute of Canada. Has twice chaired the prestigious AIA Gold Medal/Firm Award advisory jury.

P134

P44 Foster+Partners

Is an international studio for architecture, engineering and design, led by Founder and Chairman Norman Foster and a Partnership Board. Founded in 1967, the practice is characterized by its integrated approach to design, bringing together the depth of resources required to take on some of the most complex projects in the world. Over the past five decades the practice has pioneered a sustainable approach to architecture and ecology through a strikingly wide range of work, from urban masterplans, public infrastructure, airports, civic and cultural buildings, offices and workplaces to private houses and product design. The studio has established an international reputation with buildings such as the world's largest airport terminal at Beijing, Swiss Re's London Headquarters, Hearst Headquarters in New York, Millau Viaduct in France, the German Parliament in the Reichstag, Berlin, The Great Court at London's British Museum, Headquarters' for HSBC in Hong Kong and London, and Commerzbank Headquarters in Frankfurt.

P120 McGregor Coxall

Is a multi-disciplinary design firm located in Australia, China and the UK dedicated to assisting cities to achieve sustainable prosperity. CEO Adrian McGregor founded the firm in 1998 and was joined by Philip Coxall in 2000. The two studied landscape architecture together at the University of Canberra in the mid-1980's. Their international team provides services through Landscape Architecture, Urbanism and Environment disciplines. Embracing leading digital technologies, the firm delivers design solutions for complex urban and environmental challenges. Biocity Research was established in 2006 to enable partnerships with universities and scientific agencies.

P162 BIG

Was founded in 2005 by Bjarke Ingels, is a Copenhagen, New York and London based group of architects, designers, urbanists, landscape professionals, interior and product designers, researchers, and inventors. Currently involves in a large number of projects throughout Europe, North America, Asia, and the Middle East. Believes that in order to deal with today's challenges, architecture can profitably move into a field that has been largely unexplored. A pragmatic utopian architecture that steers clear of the petrifying pragmatism of boring boxes and the naive utopian ideas of digital formalism. Like a form of programmatic alchemist, creates architecture by mixing conventional ingredients such as living, leisure, working, parking and shopping. By hitting the fertile overlap between pragmatic and utopia, once again finds the freedom to change the surface of our planet, to better fit contemporary life forms.

P120 CHROFI

Is an Australian architectural practice based in Sydney, founded in 2000. Is currently directed by John Choi, Tai Ropiha and Steven Fighera. Founding partner, Tai Ropiha received his B.Arch in 1988 from the University of Adelaide. Is a registered Architect since 1990. Periodically teaches at University of Technology Sydney, University of NSW, Masters Class Design Studio and University of Sydney. Senior Designer, Susanne Pollmann received her B.Arch with distinction in 1995 from the Iowa State University and M.Arch in 1998 from the Yale University School of Architecture. Joined CHROFI in 2014. Associate, Joshua Zoeller received his B.Arch in 2008 and M.Arch in 2011 from the University of NSW. Joined CHROFI in 2009 and became a registered Architect in 2015.

Michèle Woodger

Is a writer based in London. Studied as an undergraduate at the University of Cambridge and as a postgraduate at UCL. Was previously the editor of a market-leading construction publication and website, and has worked in architecture publishing for ten years. Has recently been awarded research funding from the RIBA Gordon Rickett's Trust and the SAHGB Society of Architectural Historians of Great Britain to study public lettering in London's built environment. Has written for the RIBA Journal and the Lettering and Commemorative Arts Trust.

P28 **FOS** Foundry of Space

Is a Bangkok-based design office practicing architecture and urbanism. Position themselves at the convergence between architects and cultural analysts where socio-economic, political, environmental and other relevant factors are among our key design parameters.

Makakrai Suthadarat received his B.Arch with honours from Silpakorn University in 1999 and worked in Architecture Division, Dublin City Council, the Republic of Ireland between 2000-2001. Subsequently he moved to London to continue his study in Design Research Lab [DRL], AA School for his Master of Architecture and Urbanism. After graduation, he then worked for Zaha Hadid Architects between 2004-2006. In 2010, he founded FOS. He has been a visiting critic and lecturer at several universities in Thailand and Singapore. Their projects were selected as the finalist in WAF World Architecture Festival in 2014 as well as 2018, and also has been selected as the 2018 Architizer A+Awards Popular Choice Winner in the Shopping Center category.

©Miguel Galiano

P144 **Dominique Perrault Architecture**

Was born in Clermont-Ferrand, France in 1953 and graduated in architecture from the ENSBA Ecole Nationale Supérieure des Beaux-arts, Paris in 1978. Received his Postgraduate degree in History from the EHESS École des Hautes Études en Sciences Sociales, Paris in 1980. Founded DPA Studio in Paris (1981), Berlin (1992), Luxembourg (2000), Madrid (2006), and in Geneva (2013). Was the Curator of the French Pavilion for the 12th Venice Architecture Biennale in 2010. Has been awarded the Officer of the French Legion of Honour in 2012. Was a Praemium Imperiale laureate for architecture in 2015. Has been a Member of the Académie des Beaux Arts in the architecture section since 2015. Received the Equerre d'Argent Prize in 1990, Mies van der Rohe Pavilion Award in 1997, World Architecture Award in 2001 for best industrial building and in 2002 for best public building, Building of the Year Award by Archdaily in 2017. Has taught at ETH Zürich in 2000-2003 and at EPF Lausanne in 2013-2018.

P176 **ROSSETTI**

Was founded in 1969, ROSSETTI is a family-owned firm of nearly 100 architecture and planning professionals with a global focus on sports, entertainment, commercial, hospitality, and interiors. 'Designing Experiences that Generate Value' as their philosophy, creates spaces that engage, stir emotions and imprint memories previously unthinkable.

Matt Rossetti studied sculpture at University of Florence, and received Bachelor of Science in Architecture and Master of Architecture at University of Michigan. He became president of ROSSETTI in 1999 and has grown the 48-year-old company into an internationally-recognized design and planning firm. He leads the firm's framework for innovative thinking, believing that architecture must have a third dimension beyond form and function. Jon Disbrow, completed Bachelor of Science In Architecture at Illinois Institute of Technology and Master of Architecture at Cranbrook Academy of Art, is an award-winning designer and accomplished technical architect. His projects span from macro to micro, including sports, hospitality, residential, mixed-use, and civic projects. As a technical architect at ROSSETTI, he relies on his comprehensive knowledge of construction types and methodologies to implement design concepts and solve the firm's most challenging technical projects.

图书在版编目(CIP)数据

建筑的公共与私域 : 汉英对照 / 英国福斯特建筑事务所等编 ; 于风军等译. — 大连 : 大连理工大学出版社, 2019.10
(建筑立场系列丛书)
ISBN 978-7-5685-2224-3

Ⅰ. ①建… Ⅱ. ①英… ②于… Ⅲ. ①公共建筑－建筑设计－汉、英 Ⅳ. ①TU242

中国版本图书馆CIP数据核字(2019)第221655号

出版发行：大连理工大学出版社
（地址：大连市软件园路80号　　邮编：116023）
印　　刷：上海锦良印刷厂有限公司
幅面尺寸：225mm×300mm
印　　张：13.5
出版时间：2019年10月第1版
印刷时间：2019年10月第1次印刷
出 版 人：金英伟
统　　筹：房　磊
责任编辑：杨　丹
封面设计：王志峰
责任校对：张昕焱
书　　号：978-7-5685-2224-3
定　　价：258.00元

发　行：0411-84708842
传　真：0411-84701466
E-mail：12282980@qq.com
URL：http://dutp.dlut.edu.cn

本书如有印装质量问题，请与我社发行部联系更换。

建筑立场系列丛书 01：
墙体设计
ISBN：978-7-5611-6353-5
定价：150.00 元

建筑立场系列丛书 02：
新公共空间与私人住宅
ISBN：978-7-5611-6354-2
定价：150.00 元

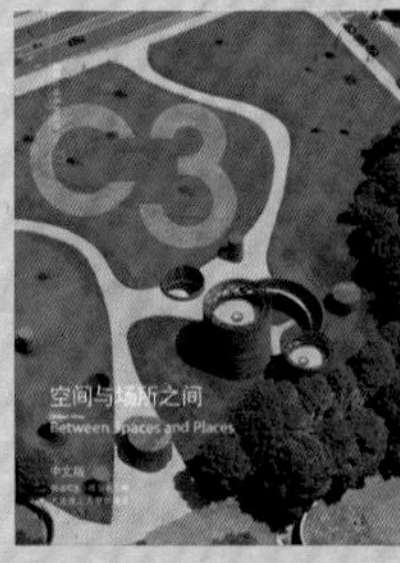
建筑立场系列丛书 10：
空间与场所之间
ISBN：978-7-5611-6650-5
定价：180.00 元

建筑立场系列丛书 11：
文化与公共建筑
ISBN：978-7-5611-6746-5
定价：160.00 元

建筑立场系列丛书 19：
建筑入景
ISBN：978-7-5611-7306-0
定价：228.00 元

建筑立场系列丛书 20：
新医疗建筑
ISBN：978-7-5611-7328-2
定价：228.00 元

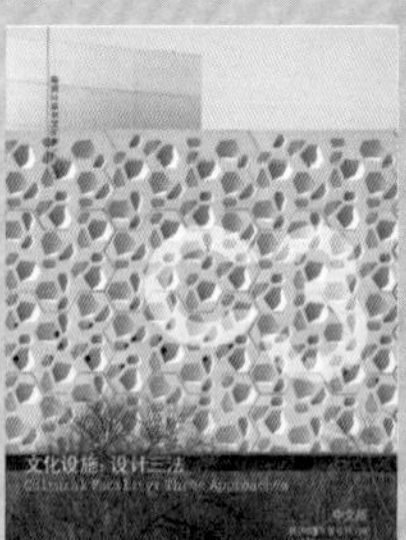
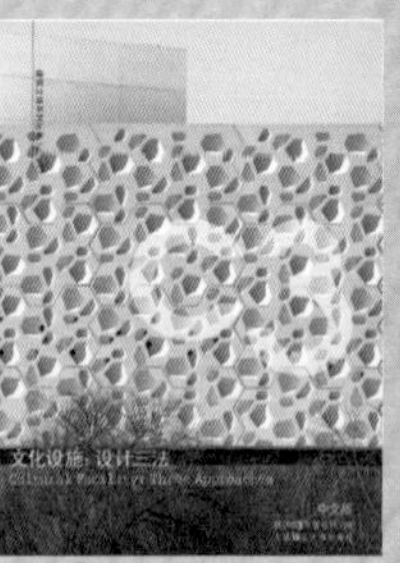
建筑立场系列丛书 28：
文化设施：设计三法
ISBN：978-7-5611-7893-5
定价：228.00 元

建筑立场系列丛书 29：
终结的建筑
ISBN：978-7-5611-8032-7
定价：228.00 元

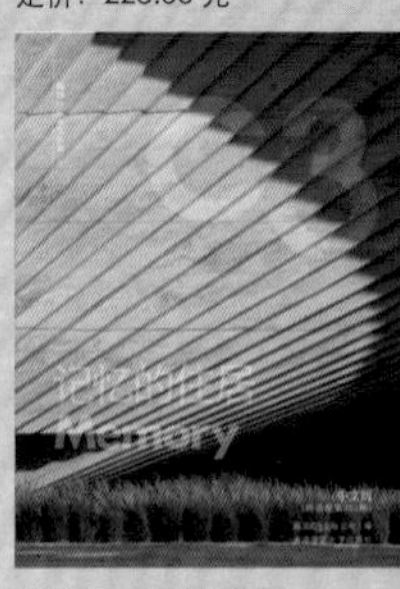

建筑立场系列丛书 37：
记忆的住居
ISBN：978-7-5611-9027-2
定价：228.00 元

建筑立场系列丛书 38：
场地、美学和纪念性建筑
ISBN：978-7-5611-9095-1
定价：228.00 元

建筑立场系列丛书 46：
重塑建筑的地域性
ISBN：978-7-5611-9638-0
定价：228.00 元

建筑立场系列丛书 47：
传统与现代
ISBN：978-7-5611-9723-3
定价：228.00 元